D1702908

The Systems Thinker – Analytical Skills

Level Up Your Decision Making, Problem Solving, and Deduction Skills. Notice The Details Others Miss.

Written By Albert Rutherford

First Printing, 2019.

Printed in the United States of America

Published by Kindle Direct Publishing

Email: albertrutherfordbooks@gmail.com

Website: www.albertrutherford.com

If you wish to receive notifications from me twice a month about:

- new book releases,

- new cognitive discoveries I made,

- book recommendations on how to develop your thinking toolkit further,

- ideas I'm pondering on,

visit www.albertrutherford.com and fill in the subscribe box.

Thank you! Talk to You soon.

Table of Contents

Introduction

I recently received a spreadsheet of 15,000 names, email addresses, GPAs, and other information for students interested in pursuing graduate degrees in a specific major at a highly competitive university. I needed to somehow select ten to fifteen of these individuals to invite to campus for a recruitment exercise. The task was certainly daunting, and the extreme amount of information in the spreadsheet made it seem impossible. That is until I started considering what I was truly looking for in these potential students. Developing a set of criteria for ideal applicants allowed me to quickly and efficiently select such a small subset from the thousands of potentials.

I first cut all applicants who weren't interested in my specific program as a first choice, and this action itself purged many individuals from consideration who were more likely to choose other programs should they get an offer in their top interest area. I then also reasoned that if I wanted the individuals invited to start graduate school in the following year, the students would all need to graduate from current programs prior to that date, and again, I saw a significant decrease in numbers on the spreadsheet. I had already narrowed down a list of 15,000 to hundreds by excluding everyone who didn't meet these two criteria.

I was able to further narrow the group of applicants by filtering the remaining individuals using criteria that were less important but allowed me to reasonably identify applicants my biggest competitors would also find highly desirable. I removed everyone from my list who didn't have at least a 3.75 GPA, were within reasonable driving distance as I was looking to specifically target students who would be less likely to be able to visit the program on their own, and participants who had attended schools with rigorous undergraduate programs, as these students were more likely to be successful in my program due to its difficulty.

The final criterion I used to gather my list of finalists was then chosen by examining those individuals remaining and areas where my program could improve. The program I am recruiting for is an area that needs diversity in sex and ethnicity. I carefully and purposefully made sure the finalists were inclusive of these groups. In the end, what had first seemed like an impossible task or one that would at least take days, turned out taking less than half an hour to complete.

Analytical thinkers are able to pinpoint and define problems, find relevant data, and create solutions that work. Analytical thinkers also test their solutions to verify they are working and correcting the identified problems or issues.

Everyone approaches problems differently. Some individuals would have been completely overwhelmed trying to select the best candidates to invite to campus from such a large number of people, especially because those selected were 1/10th of a percent of the original list. Some individuals might have considered only GPA or targeted those who would be more likely to be familiar with the university or graduate program. How an individual approaches problems can indicate whether they are an analytical or intuitive thinker, right?

Not quite, most people have the ability to think both analytically and with emotions[1]. Dr. Gordon Pennycock, an assistant professor in cognitive psychology, states that all humans are intuitive thinkers. Intuitive thinkers use their emotions or ‘gut feelings’ to make decisions. Considering human evolution, this is likely due to the reliance on instinct for survival.

Early man needed instinct to tell him when something wasn't quite right or his life was at risk. Overriding intuition might have led to exposing fatal vulnerabilities, and when fleeing a predator, relying on instinct allowed for lightning-fast decision making toward the best possible escape route. Man relied on intuitive thinking skills on a daily basis, and it's still how we make many decisions today.

Have you ever been stumped by what seemed to be a ridiculous problem? Something as simple as trying to decide between which color of the same shirt to purchase can lead to what's known as 'paralysis by analysis.'[1]

The blue shirt matches your eyes, but the green shirt goes better with your current wardrobe. Using instinct to make this decision allows you to select which color you prefer quickly versus getting stuck in the dressing room for hours, holding each garment up to the mirror over and over again unable to decide which color is the best choice.

But not every decision we make can be made on instinct. There are times when we need to employ analytical thinking skills to solve problems. Most employers today are actively looking to hire analytical thinkers. They want individuals who can identify the problems faced by the company and find creative solutions to them. Some areas where analytical thinking is a must include budgeting; making investments in the stock market, retirement plans, or other financial decisions; and even selecting which car out of several is the best choice for your needs. These are some examples of how you use analytical thinking skills in your personal life, but you're still likely to find many other opportunities to think analytically at work and home.

What is this book about?

The first installation of the "The Systems Thinker" series talks in detail about what systems thinking is, how can we broaden our knowledge about systems, and where can we use this type of thinking in practice. The second book, "The Systems Thinker – Analytical Skills" will teach you to develop analytical thinking skills. Learn how to use it to solve problems. Find creative and practical solutions to complex problems.

What are analytical skills?

It's important to note that analytical skills are just that, skills. This means analytical thinking is something you can practice and develop and is not simply a trait you are either born with or without. While having a natural interest or strength toward analytical thinking can make these skills easier to learn or feel like they're second nature, it's not a requirement. Everyone can become an analytical thinker, and everyone can benefit from the skills they will learn in this book.

Analytical thinking skills are, simply put, problem-solving skills. They are characteristics and abilities that allow you to approach problems in a logical, rational manner in an effort to sort out the best solutions.

We use these skills every day, and probably at times when we don't even think about it. Think about tasks as simple as riding a bike and then something much more complicated, such as carpentry. Both of these are analytical skills. Learning to ride a bike, while it may not seem analytical, actually takes quite a lot of continual analysis. Gaining balance to keep the bike upright while simultaneously pushing the pedals with your feet requires a basic knowledge of physics. This is as simple as leaning into the turn or curve as you change direction or maintaining the ability to stay upright if you lean too far in the opposite direction. Riding a bike also requires you to continually watch the path in front of you for debris, other bike riders or pedestrians, or potholes so that you can avoid them, and consequently, injuring yourself or others[2]. As a biker, you are constantly gathering data during the ride and then making adjustments to your behavior as you move forward.

Carpentry is also an analytical skill. It takes time and patience to learn, and carpenters go to trade school to learn these skills. They learn about proper construction techniques, reading blueprints, and properly using tools. A carpenter will think about using hardwoods for furniture that is used on a daily basis, as these types of wood are more durable and less likely to nick or scratch. The carpenter may also think about employing reinforced hardware on items such as a wooden swing set because he is thinking of the safety of the children using the item and doesn't want to risk a nail or small screw coming loose and causing injury. These are just examples of how carpenters think analytically and come up with solutions that work in his or her profession.

Benefits of Analytical Thinking

The benefits of analytical thinking are innumerable. One of the biggest assets possessed by analytical thinkers is communication. The ability to communicate well with others is a highly sought commodity both professionally and personally. Without the ability to communicate, we cannot effectively make our thoughts and ideas heard[3]. Highly effective communicators are able to conquer one of the biggest challenges faced by millennials today in both written and verbal forms.

Analytical skills help people manage conflict, consider tough topics from an unbiased perspective, or at least acknowledge they have personal bias and account for it, conduct their own independent research, and be confident in the conclusions drawn from that research[4]. These skills can help the average person make an informed decision on which candidate to vote for in elections, consider the arguments of a friend or loved one in a disagreement, and understand that humans are infallible. Analytical skills are so important because they help you meet challenges in all aspects of your life and find solutions that are effective[3]. These are key components to living a happy and informed life in all the realms individuals manage and navigate.

Costs of Not Thinking Analytically

Have you ever heard the phrases "follow your heart" or "think with your head, not your heart?" This first is an example of favoring emotional, or intuitive, thinking over analytical thinking. Ideally, individuals can utilize both intuitive and analytical types of thinking, and we refer to these people as "holistic thinkers."[5] Using holistic thinking allows the individual to move past a right or wrong/ or lose/win approach to problem-solving, and combine both analysis and feeling to develop well-rounded, comprehensive answers to complicated issues.

Truthfully, neither analytical nor intuitive thinking is better than the other. Both types of thinking carry pros and cons, and individuals should consider the situation at hand before deciding which type of thinking would be useful, or even if one is more appropriate than the other. Don't forget the innate value of the holistic approach. For example, employing analytical thinking strategies would not be the best course of action when selecting a spouse or romantic partner. This is a decision that is instinctual and based on factors like attraction, morals and character, and an appreciation for similarity. Intuitively, there is a much less likely chance a person will select a partner who they find physically unappealing, has values and morals that do not gel with their own, or who has a majority of different interests. An attempt to do so is likely to end up in heartbreak or dissatisfaction in a person's choice of romantic partner.

But there are undeniably times when reining in your emotions can lead to a level-headed and well-reasoned course of action. Analytical thinkers focus on the small picture, objective information, and use linear logic[5]. An example might be deciding what investments to make. Making smart financial decisions requires a person to consider factors such as risk, time to maturation, and previous patterns of similar investments over a period of time. In a perfect world, all decisions could be made with a holistic approach, and individuals would be confident they were making the absolute best decision emotionally and rationally.

Unfortunately, this is simply impractical. Decision makers need to learn when to allow their heart to control their head, their brain to take the lead, or when both of these powerful forces can work cohesively for the best. To learn when to employ different approaches is to help you avoid being taken advantage of, or brainwashed, or hoodwinked. Think about how many people fall for phishing scams like the Nigerian prince scam. The individuals who fall for these types of scams are typically pushing their analytical thinking skills to the back of their brains while focusing on the emotional high of spending millions of dollars in easy money.

Characteristics of an Analytical Person

- Tackle issues based on facts and logic rather than emotions.
- Shows positive performance in highly organized situations where the whole picture can be visualized. This reduces the chances of being wrong or not seeing the bigger picture.
- Has high-performance standards in problem-solving tasks, particularly when knowledgeable about the topic at hand.
- Cautious when working with strangers or acquaintances.
- Problem-solving is a strength.

- May be quiet or appear to be unemotional to others.
- Can appear cold, noncommunicative, and aloof in interpersonal relationships.
- Takes problem-solving approach in order to 'fix' the issue.
- Wants things to be rational and well organized. Does not like emotional or disorganized strategies.
- Collects facts and data before making a decision.
- Uses skills as a problem solver as the foundation for relationships.
- Hates to be wrong and avoids it like the plague. I once saw a shirt that read "I thought I was wrong once, but it turns out I was mistaken." This is the analytical thinkers' motto.
- Are often in demand for their opinions, exactness, and data-oriented expertise.
- Has a theory for almost everything.

Chapter 1: Analytical Thinking

What Exactly are Analytical Skills?

Analytical skills are the ability to apply logic to the collection and subsequent analysis of data. Those individuals that use analytical thinking skills, are able to visualize, express, and solve concepts, both concrete and abstract, using analytical thinking skills. In turn, these same individuals are to examine all the available choices and options and choose those that make sense based on the available information.

Why Analyze Information?

There are countless reasons to analyze information. Employing effective analytical decision making when selecting political candidates we want to support can have a lasting and radiating effect on legislation at the municipal, state, and national levels. Numerous individuals actively participate in national elections, i.e. presidential elections, while simultaneously ignoring the impact local elections can have on policies that affect them in their day-to-day life.

There's also no denying the impact of our economic decisions. Decisions that are made thoughtlessly may see us taking on large amounts of unsecured debt, and that, in turn, may have a detrimental impact of how credit companies operate but also view you as a potential risk. In the early 2000s, hundreds of thousands of students applied for and accepted private student loans. Due to both predatory lending practices and the inability of students to repay these large debts, the private banks administering federal loans have all but disappeared. Furthermore, the United States government has severely cracked down on exactly what credits they will fund at both traditional institutions of higher learning and non-profit institutions. All of these changes have had the intention of helping Americans avoid crippling student loan debt, but there's no denying these policies have also crippled student choice and the ability to approach higher education as exploratory.

There's also the reality of understanding how our choice affects our own relationships, especially with friends, family, and coworkers. Poor decision making at work may lead to a lack of trust and confidence with coworkers. You may find yourself routinely asked to justify your decisions. But it's also important to realize the impact analytical thinking has in our personal lives. Individuals who are not budgeting using analytical skill may find themselves in debt and a partner who is angry or frustrated over how this situation came to be in the first place. Other areas where emotional decision making may get you into trouble could include something as simple as a group of friends visiting a bar. Your friends can quickly become uncomfortable going out with you if they routinely have to clean up the messes of your poor choices due to alcohol or other factors.

Analytical skills in the workplace have endless opportunities to enhance your performance, lead to promotions, and make you one of your company's most highly sought professionals in your field. Tackling complicated problems based on data, an understanding of where your company is now and where it wants to go, and the utilization of innovative solutions are all analytical skills that make you a commodity. When reviewing resumes and performance, analytical skills are some of the most actively and fiercely pursued by hiring managers and human resources professionals. Analytical thinkers are able to pinpoint and define problems, find relevant data, and create solutions that work. Analytical thinkers help employers stay ahead of their competition and stay relevant in an ever-changing, dynamic, and professional world. Analytical skills have become the threshold job applicants are measured by and this is unlikely to change in the near future.

The Problem-Solving Process

Problem-solving is a skill everyone uses, but that doesn't mean there is no rhyme or reason to approaching the problem-solving process. According to Dr. Minwir Al-Shammari, a professor of management at the University of Bahrain, there are five important stages to problem-solving: problem formulation, finding alternatives, analysis of alternatives, alternative selection, and implementation of the selected action[6]. Together, we will explore each stage of the process and the characteristics associated with each one.

Problem Formulation

This phase is simply identifying the problem and is considered by many to be the most significant aspect when it comes to approaching problems and possible solutions[6]. It is virtually impossible to solve problems and find effective solutions if you cannot even identify what the problem is in the first place. At this step, it is vital to approach the issue with an air of skepticism. One of the first things you want to remember is not to take information as an automatic truth. Most people have heard the phrases 'Don't believe everything you read' or 'Don't believe everything you see on television or the internet.' This stage simply underlines that sentiment and cautions you to take information with a grain of salt.

Finding Alternatives

There is rarely one way to approach a problem, and now we need to ask ourselves some strategic and important questions as we consider viable solutions. One of the questions we need to ask is 'What are the viable solutions?' This is an important consideration because we need to be open-minded when considering the full spectrum of possible solutions. Keep in mind that solutions can be complicated, involve multiple facets, and potentially be costly. There is rarely a cheap and easy solution to significant problems[6]. When weighing and measuring all of these options, that is when costs and contributions should be evaluated. An extremely expensive solution that makes minimal impact would likely not be worth the expense. However, an equally expensive alternative that solves the majority or most important aspects of an issue can be money well invested. The least important considerations when looking at viable solutions are limitations and constraints. These two aspects should only be considered toward the end of the problem-solving process[6].

Analysis of Alternatives

Analyzing the alternatives you develop in stage two can be agonizing and grueling. When making a decision you have a number of influences that will affect what you decide. You may use quantitative methods to logically and dispassionately make decisions. or you may rely more on your judgment, experience, and intuition[6]. Quantitatively examining an issue can be highly effective in making difficult decisions. For example, let's pretend you are a supervisor at work and one of your employees isn't performing to your standards. Internally you may be struggling with the impact this person may face from an employment termination but reviewing hard data can be an eye-opener when reviewing the impact an underperforming staff member has on your company. For example, you may be emotionally caught up in how this person will support his or her family or how much you like them as a person, but when you look at facts, such as poor customer ratings or excessive absences, you have an undeniable dataset that shows you how the employee is putting the entire company or division at risk.

If you rely exclusively on intuition to make decisions, you may find the solutions you select are ineffective. For example, let's say you are in the market for a new car. If you go to the dealership with little concept in mind of what your needs and wants are, it's extremely possible you could find yourself the owner of an impractical new car. The last time I purchased a car I really thought about all the features I needed in a car to make it the ideal automobile for my family. I needed a car with ample cargo space for transporting my pets, a third row so friends and family didn't need to plan costly personal transportation during extended visits as I don't live near any of them by a considerable distance, and a car that was very reliable. I didn't want to have to take my car to a mechanic on a regular basis, and I wanted a car that would last, or as I say drive until dies.

All that being said, there is no denying emotions and reputation played into my final decision. The brand of the car I selected routinely has a very high reputation with both customers and industry researchers with no personal stake in ratings and reviews. I felt comfortable selecting the make and model car I chose because I knew it should last several hundred thousand miles. I also selected a car that had one of the largest cargo areas on the market. Learning this information, and even the make and model of all the cars I considered, took research and evaluation on my part. It's not completely unheard of for individuals to visit a car dealership only to end up purchasing a sports car that doesn't meet the needs of a family man or a minivan that is entirely too much automobile for a single person or childless couple. Analyzing options and taking the best of both analytical and intuitive thinking can help you make an effective decision you can be happy with.

Selection of an Alternative

Contrary to the heading of this section, the appropriate solution isn't always cut and dry, because the preferable solution isn't always obvious. This is when using the quantitative and qualitative methodology is key to determining the best possible solution. The previous section should have given you an idea of how to combine intuitive and analytical thinking to sift through the viable options and find the best ones for the scenario at hand.

Implementation of the Selected Action

Implementing a chosen course of action isn't something to be taken lightly. Plans and actions, despite the intent for success, can easily be derailed due to poor or hasty implementation[6]. There are countless factors that influence effective implementation including: interpersonal dynamics, the behavior of the organization, and communication styles just to name a few[6]. There are plenty of other factors to consider during implementation depending on the setting of the problem.

Analytical Thinking Skills and Dimensions

Developing and implementing analytical thinking skills may seem technical and overwhelming. You are asking yourself if you even need to worry about developing this skill set. Unequivocally, I can tell you yes! It's easy to think of this skill set as tools to be relegated to the workplace or only used in special circumstances, but the truth is analytical thinking is used multiple times a day in areas far beyond the professional world. Below are some ways you use analytical thinking multiple times a day in your regular life.

Communication

What good are analytical skills if you lack the ability to tell others what you know?[7] As previously mentioned, effective communication is priceless. There are multiple types of communication so don't stress too much if you're really good at written communication, but struggle with the verbal aspect. Think about something like marital or family therapy. Individuals need to be able to convey to the therapist what the problems are but also listen to other participants and the therapist. If you can't do this, the ability of therapy to resolve issues is going to be limited.

Creativity

Very few people have the ability to single-handedly contemplate all the viable solutions to any given problem. While the phrase 'think outside the box' has become a bit of a cliché in our culture, this is exactly what is meant by creative solutions. It's also important to remember that finding creative solutions often requires collaboration between multiple individuals[7]. This is completely normal and should be embraced during the analytical thinking process. Never underestimate the value of multiple minds during the development of planning, improvements, or restructuring. Collaboration alone is a kind of creativity, and don't sell yourself short by relying exclusively on yourself for all things analytical thinking.

Critical Thinking

This book is about analytical thinking and developing the skills to help you become an excellent analytical thinker. However, I would be remiss if I ignored the role of critical thinking. Critical thinking is vital to developing strong analytical skills, the skills that allow you to evaluate information and make decisions based on that information rather than focus on the emotions or judgment that may otherwise influence decisions.

An example would be recent changes I have implemented with master's students in my graduate program. Until recently, master's students were not placed on probation in my department, even due to poor academic performance. As a result, I have five planned graduates who are in very real danger of not graduating due to poor grades. I also have a student who needed to be dismissed at the conclusion of the previous term for the same reason. Critically, these underperforming students should have been dismissed from the program based on their inability to complete the degree requirements regardless of their reasons for their performance.

As an analytical thinker, I considered each student's personal circumstances or other factors that may have led to an undesirable performance, such as English as a second language and the student's ability to grasp the material. I use my judgment, experience, and intuition in addition to the data I pulled on each student to determine who should be dismissed versus those who may need an additional term to recover academically.

Data Analysis

Data analysis is the ability to examine datasets and find trends in that data. This is more than just reading and understanding information. Data analysis is making sense of that information and picking up on the trends within it[7]. Think back to the new car situation we previously reviewed. Beyond selecting an automobile that fits your needs and wants, you need to think about the overall cost of the car. Data analysis is one of the skills that help you determine this information and evaluate what you can expect to spend on a car beyond the sticker price. Overall cost typically considers expenses such as insurance, gas mileage, and financing rates for those who aren't able to pay cash for their vehicle. Depending on certain factors, a $20,000 car may cost closer to $30,000 because of these variables. Analytical thinkers will consider all of these factors before making a decision.

Research

I personally consider research second to communication in the analytical thinking skill set. If I could list research ten times, I would.

Conducting unbiased, independent research is one of the most valuable tools in your analytical skill set.

I recall when we were expecting our first child. I researched, researched, and researched some more when selecting appropriate items for my young one. One of the biggest issues I refused to waiver on was the selection of our child's car seat. As an inexperienced father-to-be, all I cared about was my child's safety. I researched safety ratings and accident performance on all major car seat manufacturers and models. In the end, I selected 'the best.' The best in terms of safety and performance was worth every penny to me as a dad who knew children weren't replaceable. I continued to prioritize, investigate options, and collect data on infant and toddler items for years. This led to selecting top-rated toddler boosters as my children grew or selecting a balance bike versus one with training wheels because my research showed it helped children learn bike riding in a different way. Many people have hobbies and passions and devote time, effort, and analytical skills when selecting items that serve them in their hobby. They want the best items for their needs and analytical skills are vital to choosing these specific things.

How to Improve your Analytical Skills

Successful businessman, Warren Buffet, listed analytical skills as the most vital skills for young people to be successful[8]. It's hard to imagine analytical skills being the key components to success, especially when a person considers investments, financial savvy, or political savvy. It's easy to chalk up success, especially not fully understood success, to who you know or excellence in your chosen profession, but don't be fooled or downplay the importance of analytical thinking.

Ask the right questions

Asking the right questions can mean the difference between solving a specific problem and solving a problem that doesn't address the issue at hand. Some key questions include 'What do we already know,' 'What do we need to know,' and 'When is this information needed.' Without addressing these vital questions, it is easy to become sidetracked and address questions that don't pertain the major issues at hand. Some other relevant questions include asking yourself about the expected results, the planned response when we have the results, and how much time is permitted to conduct the research. A timeline for conducting research is particularly crucial when acknowledging your lack of knowledge on a particular topic and planning a course of action for acquiring it.

Realize What You Don't Know

There are very few people in this world, if any, who are all knowing. Don't get bogged down by your lack of knowledge. Realize you're in good company amongst your peers and make adjustments to account for this limitation. One of the most important rules is to remember that you don't need to be embarrassed to ask simple questions. Sometimes you need clarification, or what may seem simple on the surface can easily turn out to be very complicated. It's also imperative to start with the basics to ensure you account for all your bases and remember to let your research be your guide. When you influence the direction and pattern of your research, you're less likely to get unbiased and accurate results.

Don't Make Assumptions

Have you heard the saying making assumptions makes an ASS out of U and Me? You don't want to get caught in this crude, but valuable saying. You may need to ask yourself and your client if they really know what they want. It's not uncommon for people to examine specific subsets of information on request, but the more you talk to a requestor, such as a supervisor or client, and delve into the information that was asked for, to realize what was asked for isn't really the information truly needed to address this problem. This is why not making assumptions is so vital to effective problem-solving.

Don't forget to do your own research. It's completely logical that the requestor isn't the first person to face the issues causing problems. Ask yourself if others have gathered the same information and how long it will take you to uncover the information. This is key information to consider during information acquisition, and don't sell yourself and skills short.

Don't Take Things at Face Value

Remember the previous section about not believing everything you read? That word of advice stays true but remember that not all incorrect data is presented with a malicious intent. Sometimes, people simply make errors. People aren't infallible and neither are machines. Have you ever gone shopping at an outlet mall? Most of the merchandise for sale in the stores at these malls have defects. Maybe the company's logo isn't properly centered, or the quality of the goods isn't what you would expect in a retail store. Outlet malls are the perfect example of machine error.

These errors aren't limited to retail. Remember to use your analytical skills to double check information presented in everything from internet articles to professional journals. People who are professional researchers or journalists aren't above human error or even intentional deception. For several years, Dr. Andrew Wakefield actively promoted that the measles, mumps, and rubella vaccine had a strong correlation to autism. Despite the fact that correlation doesn't equal causation, Dr. Wakefield strongly suggested this particular vaccine did cause autism in some children who received it. Medical researchers who attempted to recreate Dr. Wakefield's studies and results were completely unable to do so. Sadly, it came to light that Dr. Wakefield was altering his results because he wanted to promote his own measles vaccine, which would have resulted in significant payouts for Dr. Wakefield. This example is just one of many reasons why you should cross-check your statistics and research; critically evaluate the information and make decisions on its value; and actively question the credibility of sources, especially online sources.

Turn Information into Knowledge

So now that you have all this data, what do you do with it? One of the easiest ways to answer this question is to keep your client or requestor in mind. If you're considering only your thoughts and goals, you may completely miss the goals of the person or organization needing the information you are researching. There is nothing wrong with adding your perspective to the work you've done and communicating that to the person who requested the information for consideration or not.

It's also imperative that you categorize and facilitate your organizational methods as you maneuver the analytical process. Some key points to keep in mind are: examining information for patterns, summarizing the information you've collected, and assembling the information in a way that is easy and efficient to review.

Key Takeaways

- Problem-solving isn't an easy process. There are multiple facets to effective problem-solving.
- Analytical skills serve you in both your personal and professional life. Developing strong analytical skills helps you to be more critical of the information you are presented within all areas of your life.
- Analytical skills can be practiced and improved. If your analytical skills are

weak or need improvement, that’s not a black mark, and you can still be a successful analytical thinker.

Chapter 2: Analytical Thinking Techniques

How is Analytical Thinking Built?

It takes a lot of flexibility to cope with the challenges and pressures of problems as they pop up in your day-to-day life. One of the most effective strategies for handling issues is to have a strategy for approaching problems. Analytical thinkers use a scientific approach to problem-solving[9]. Below is an illustration of the analytical thinking cycle[9].

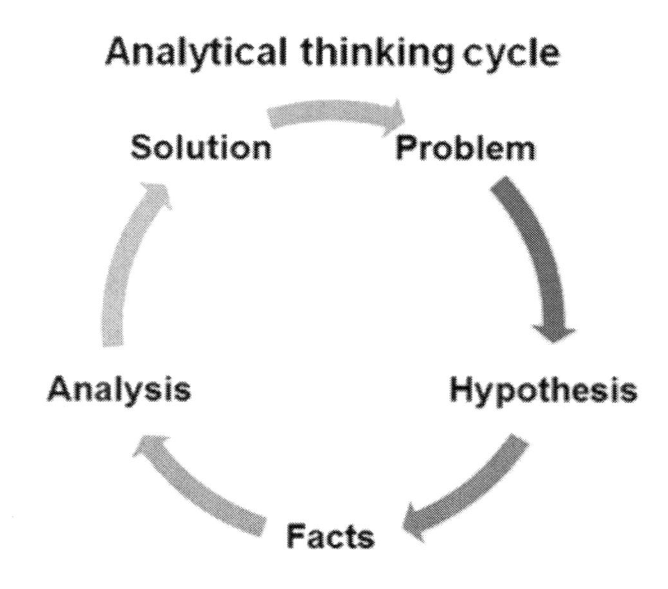

The first step to approaching any problem is defining that problem. Is your car breaking down excessively? Are you constantly arguing with a loved one over something like money or politics? Are you unhappy in your career? All of these are problems that an analytical thinker needs to identify. You can't tackle problems and develop viable solutions if you can't even pinpoint the problem.

Techniques for identifying problems include strategies such as asking yourself 'why' five times. Each time you ask yourself why the problem exists, you may discover differing reasons, and this, in turn, can help you find the true cause of your problem[9]. Asking yourself 'who, what, where, why, and how' can be some of the most effective questions you ask when zeroing in on a problem.

Also, consider looking at cause and effect. For example, every time you bring up a certain topic, like politics or religion, does this lead to an argument with a family member that results in estrangement? Knowing the cause—sensitive topics—can help you see where your relationship might be going wrong. A similar approach can be taken with nearly every problem. If you're dissatisfied with your career, use this method. If you feel underappreciated or dispassionate toward your career, you may want to really examine what leads to these feelings. Does your boss fail to acknowledge your contributions, or do you feel that another field altogether would bring you more happiness and satisfaction? Both of these are causes that could lead to your unhappiness at your job. Ask your numerous questions and remember there are multiple approaches to discovering the root cause of a problem.

After identifying the problem, you want to develop a hypothesis, or even more than one, to help you develop a solution that works. It's not uncommon for complex problems to have complex causes. Establishing your hypotheses helps you target what data you need to collect[9]. The pivotal component to solving your issue is drafting an inclusive list of all the possible issues creating the problem. From there, you want to strategically reduce duplicate and overlapping issues in an effort to draft your hypotheses. Remember not to make your questions too broad or too narrow and don't develop so many hypotheses you can't remember them all.

Developing two to five key questions that are well thought out and reasoned will help you far more than ten questions that lead to a lot of unnecessary data collection[9].

Speaking of data collection, now that you've developed your hypotheses, it's time to go on a scavenger hunt to collect the facts. Factual data collection is an absolute requirement for the identification of workable and accurate solutions. One of the first key requirements of data collection is determining what data needs to be gathered in the first place. It seems like a bit of a no-brainer, but don't gloss over this step. Truly evaluate and determine what facts and information you need to address your problem. Remember there are also multiple kinds of data out there too.

Qualitative data basically boils down to a narrative. It is either a spoken or written response to your questions and holds some special benefits over quantitative data. Qualitative data is often referred to as 'storytelling' because it includes extensive additional information beyond a simple response to the hypothesis question. If you were trying to evaluate how divorce affected children, qualitative responses might include a description of the behavior that led up to the divorce, the ages of the children, how the children felt about each parent, and on and on. Collecting qualitative data can help you understand more facts and circumstances and how they impact the problem you're trying to solve.

Quantitative data relies on numbers and shouldn't be considered better or worse than qualitative data. Each data collection method has its benefits and there may be times when you want to use both methods of data collection. Unlike qualitative data, quantitative relies on numbers and calculations. This type of data is incredibly useful because it can help show the prevalence of a problem or even show ratings on how an issue affects others. Quantitative data is easier to evaluate and often is thought to be more impactful just for the sheer amount you can easily collect.

Let's revisit the issue of divorce and the impact it has on children. Quantitative collection would allow you to review information on hundreds of children. As a researcher you would be able to cut through information you might consider superfluous, like child/parental relationship, spousal relationships, or child age, but you would quickly and efficiently be able to evaluate if divorce has an impact on children or not. One of the best techniques you can use when considering what kind of data collection you want to use is to think about what question you want to answer. If you want to answer a question that addresses the why and how and impact, perhaps qualitative data collection is a wise choice. On the other hand, if you're looking to evaluate how many of your make and model car have been to the mechanic in the past year, a quantitative methodology would likely better suit your needs.

The next step in building analytical thinking is analyzing all this data you've collected. One of the most common types of analysis is a 'what if' analysis. This type of analysis evaluates all the potential solutions and potential outcomes. There are many ways to conduct a 'what if' analysis including storyboarding, decision trees, and simulation[9]. The goal of 'what if' analysis is to identify comprehensive repercussions and avoid those outcomes. Another type of analysis is the 'cost-benefit analysis.' This procedure is basically as described in its title. In this scenario, you would review all possible solutions, and weigh the cost of those solutions in comparison to the benefits. Costly solutions with minimal benefits should be avoided. Ideally, you would be able to develop an outcome with a minimal cost but extensive benefits. This is a dream we can hope for, but a cost-benefit analysis should be a regular part of your data evaluation.

Another approach to data analysis is a SWOT analysis. SWOT stands for Strength, Weakness, Opportunities, and Threats. This is a highly effective type of analysis because beyond identifying positives and negatives (strengths and weaknesses,) it can help you note things you should and shouldn't do[9]. Avoiding opportunities and embracing threats is significantly less likely to lead to solutions that work. There is no denying that a SWOT analysis is one of the most common tools for critical planning[9]. SWOT is somewhat subjective, but don't let that stop you from employing this vital tool. There are many types of data analysis and one type isn't necessarily better than the other. Find the analytical method that works for you and implement it, or better yet, implement multiple methods to ensure effective problem-solving.

And finally, create and implement a solution. Many of us have heard the phrase 'out of the box' in relation to solving problems[9]. Don't be afraid to develop out-of-the-box fixes to common conundrums. Implementing the previously reviewed process can help you cultivate unique ideas that solve problems that have persisted for years or decades. Remember that no solution is guaranteed, and don't forget to employ a solution rating matrix that takes into account analytical thinking to help you select the best and most effective solution.

Analytical Skills Require Practice

Were you born knowing how to play the piano or drive a car? Of course not! You have to practice to learn how to play a musical instrument or properly and safely operate a vehicle. Analytical skills aren't any different. You have to practice these skills and strengthen your analytical thinking muscle.

Herbert Simon's Decomposable Matrix

Decomposable matrices come from Herbert Simon's, an economist and political scientist, view that hierarchies are comprised of semi-independent subsystems that are less complex than the previous system[10]. Simon believed that complexity in the world evolved from simple structures organized into more formal hierarchical systems. For example, have you ever saved a file on your computer? Of course you have! Most computers have a complex hierarchy, the further you delve into the file system, the more a decomposable matrix becomes obvious. Let's imagine you need to save a file for work. Computers start out listing multiple drives. Nearly all computers have an included hard drive, but networked computers may have multiple drives for multiple purposes. Personally, I have a hard drive, a personal drive networked to an external server, and a shared drive containing files multiple people can access. Beyond this, I have folders and subfolders I use to house my documents based on topics and confidentiality. There are countless decomposable matrices within the computer drive system and even beyond.

Hierarchies and matrices aren't just decomposable or not. There are also near decomposable hierarchies and totally decomposable hierarchies. The primary difference between total decomposable hierarchies and near decomposable hierarchies has to do with interaction. Near decomposable hierarchies maintain some interdependence between the subsystems[10]. Totally decomposable hierarchies have no interaction between the subsystems[10]. Understanding these different types of hierarchies helps you analyze them by breaking them down to the subsystems and then further analyzing the components. You can conduct this analysis by following Simon's five steps: determine, list the major, construct a matrix, use a one- to five-point scale to evaluate the degree of relationship, and select the highest weighted interactions[10.]

Let's use Simon's analysis in an example. Consider the structure of law enforcement. The law enforcement structure includes a commissioner, chief, sergeants, lieutenants, detectives, and patrolmen. This isn't an exhaustive list, but it's a good listing for smaller municipalities. Let's presume the problem is corruption on the force. In this instance, we can *determine* the hierarchy has analyzable subsystems because each level is highly connected to the next. We've already listed the major subsystems and we can examine them both independently and how they interact with one another. The next step is to construct a decomposable matrix. A decomposable matrix will contain columns of the different subsystems and rows of issues that affect those subsystems. Examples of issues that lead to law enforcement corruption might include low rates of pay, lack of ethics training, and power over average citizens. The next step is to weight the analysis of the subsystems on a scale of one to five. Of the five steps of analysis, which is most important? Once we determine that information, we need to weight these interactions. For example, internal interactions between officers may be weighted less than interactions between officers and the general public. Interactions with the general public may be weighted less still, than interactions with media outlets. In this example, interactions with media outlets may be weighted most highly because they have the potential to influence citizens' opinions of law enforcement in their town.

Dimensional Analysis

Elisabeth Jensen's dimensional analysis is an analytical method to clarify and explore the scope and limits of a problem. It examines five elements of a problem: substantive dimension, spatial dimension, temporal dimension, quantitative dimension, and qualitative dimension[10]. The purpose of these dimensions is to answer the five crucial questions: what, where, when, how much, and how serious[10]. These questions are further evaluated by answering specific questions. Dimensional analysis is broken down by using the following five techniques:

- Write down the problem;
- Quickly describe the problem in terms of: What? Where? When? How much? How serious?
- Use the descriptions to address the questions in each dimension (see chart below)
- Analyze your responses to the questions by thinking about their implications and their effect on solving problems
- Choose which areas are most applicable to the problem and need additional analysis[10].

Substan-tive	**Spatial**	**Tempo-ral**	**Quanti-tative**	**Quali-tative**
Commi-ssion or Omissi-on?	Local or Distant	Long-standing or Recent?	Singular or Multiple ?	Philoso-phical or Surface?
Attitude or	Particular Loca-	Present or Impending	Many or Few	Survi-val or

Deed?	tion(s) with a Location	?	People?	Enrichment?
Ends or Means?	Isolated or Widespread ?	Constant or Ebb-and-Flow	General or Specific	Primary or Secondary?
Active or Passive?			Simple or Complex ?	What Values are Being Violated ?
Visible or Invisible?				Proper or Improper Values?

One of the struggles encountered with using this method is that there is little instruction on how to analyze a problem. One of the reasons for this is that Jensen describes this analysis as an exploratory method when analysis has become an issue. The intent is to require multiple considerations related to the problem selected to be contemplated with the idea that solving the problem may be a smoother process than if no analysis had been done at all. One of the best ways this analysis can be used is through a checklist that is used as a methodical guideline to help provide perspective during the later stages of problems.

Organized Random Search

Frank Williams' methodology to systematically analyze a problem suggests that instead of indiscriminately searching for ideas, it is more productive to break them down into parts[10]. The primary benefit of this process is that it helps create direction or generate ideas. Williams' analysis only has two primary steps:

- Inspect the problem for possible subdivisions or ways of categorizing parts of the problem
- Write down different subdivisions or parts and use them to generate ideas[10].

One of the areas Williams has implemented his analysis involved total random recall. Let's say you were required to list all the countries in the world. That's a challenging task that has stumped more than one student enrolled in a basic geography class. But what if you developed a study strategy that utilized a prographical (visual) technique. Instead of trying to memorize countries alphabetically, a student could arrange them east to west or north to south. The student could use visual means to improve his or her performance on an exam, and historically a person's performance is dramatically improved.

It's important to note as a general review that there are few, if any, interactions which can significantly benefit from the Organized Random Search method[10].

Relevance Systems

Relevance systems are a method of organizing information about a problem through successive assessments that clarify major problem components[10]. Again, this is a methodology that leads to a visual example. As the researcher reviews the information, as each component is listed, each component is connected with preceding ones. This eventually creates a pyramid-like structure. By working both downward and upward (the connections between employees and managers) the accuracy of elements involved and their relationship to one another can be quickly and easily evaluated.

According to Jim Rickards, there are two types of relevance systems: single and binary. Luckily, the explanation of the two systems is simple and easy to understand. A single system consists of all the components related to a single problem whereas a binary system interfaces with two single systems that interact across the lower levels of these two systems. Binary systems are special because they can be used to pinpoint connections between and within systems[10].

There are two ways to create a relevance system. Similar to the organized random search methodology, you can start with the elements at the top and work your way down or vice versa. Both methods are effective used by themselves, but research reports that the best possible results derive from using both methods[10].

You may be asking yourself how to build an effective relevance system. Here are some key points[10]:

- List the most important part of the problem. This is level one.
- Write out the elements that are derived from the first level. This creates level two.
- Continue writing out elements until all possible elements have been established, and the lowest level is identified. This is generally done by addressing the question, how? Note that higher level elements are determined by asking, why.
- When finished, assess validity by working upward.

- Use the lower level elements to suggest solutions to the problem.
- If the problem overlaps an additional area and should be integrated with it, build a secondary system so that its lower-level elements connect with the first system. This creates a binary relevance system.
- Investigate the connection to determine points of singular or mutual influence and consider possible factors that could influence the objectives in either system.

Rickards described how a binary relevance system could be used to enhance a company's marketing strategy. A single system is built with the highest-order element. In this instance—increasing the number of children attending Ms. Marie's ballet class. This is followed by the lowest level elements, those that address the question, how. If Ms. Marie wanted to look at her marketing plan to increase her clientele and bring in new ballet students, a secondary system could be created.

That marketing plan could then be incorporated into other decisions to increase business by selling ballet slippers and uniforms and other accessories. The resulting binary system allows Ms. Marie to analyze possible constraints such as hours of availability for classes, her ability to buy and maintain adequate stock for merchandise, and room size that might influence the solutions she comes up with as potentially viable.[10].

The evaluation of relevance systems relies on progressive breakdowns of problem elements and the development of new problem definitions. A problem is broken down so all of its major elements and connections can be evaluated. Opportunity interfaces are a crucial strength of relevance systems because when the problem solver has already considered constraints, the probability of solution revisions should be significantly reduced. In addition, lower-level elements can increase the richness of problem solutions because when multiple combinations of problem elements are considered together, rarer solutions will be an outcome[10].

Simple Tips to Increase Analytical Thinking on a Day-to-Day Basis

- Read books from more than one point of view. Most books are written from the point of view on one or two characters but try to imagine how the villain would have written the novel or what the hero's former love, whom he left behind to marry the heroine, would have to say about things. This type of activity opens your imagination, expands your mind, and is stimulating and fun all of which is critical to developing analytical thinking skills.
- If you're not much of a reader, you can also work on boosting your math skills. Calculus, algebra, and statistics rely on logic and reasoning. They all take precision and a step-by-step approach to calculate the correct answer.

- Try going for a walk and observing everything occurring around you. Leave your music and earbuds at home. The sounds around you are important too. One of my favorite activities is taking my English Springer Spaniel, Winston, on a walk with me because he is very excitable and has a high work drive. It comes out in him during walks and it's such a joy to watch that instinct of hundreds of years of breeding as a hunting dog come out. He sniffs the air and looks up. I follow his gaze. He sniffs the ground and looks at the ground and sniffs more and asks what he smells. And then he proceeds to walk me around the neighborhood for the next hour, where we sniff and investigate together.

Key Takeaways

- Analytical thinkers approach problems scientifically.

- There are numerous methodologies available for analytical thinkers. Don't think you are stuck with one approach. You can often use multiple approaches to find the best possible solutions.
- Individuals aren't born with analytical skills. They take practice and that's completely normal. What's more, these are skills you can use anytime you need to make an important decision.

Chapter 3: Synthetic, Systemic, Critical, and Creative Thinking

Analytical thinking is the quintessential tool for understanding all of the different parts of a situation. It is defined as having the capacity and skill to dissect and analyze or developing the ability to think in careful and perceptive ways in an effort to solve problems, analyze data, and remember and utilize information. However, analytical thinking doesn't stand alone. It is but one cognitive process in relation to four other concepts: synthetic, systemic, critical, and creative thinking.

Synthetic Thinking

Analysis is a considerable thinking tool when used to understand situations and all the different parts affecting that situation. When we break apart the different pieces involved in a situation it's incredibly easy to lose sight of how these parts interact with one another and within the situation, and this is how we end up in 'paralysis analysis.' This is because analysis tends to 'shade' the interactions and make them more difficult to identify. When the interactions aren't as visible, our insight is also reduced, which can cause a situation to go from not so great to a mega disaster.

The tool we use to make sense of interactions and how things work together is synthesis. Synthesis is much more than simply putting something back together after you've taken it apart for an investigation. If analytical thinking helps us understand the different parts of a situation, then synthetic thinking allows us to truly understand how these parts work together. Another example of the differences between analytical and synthetic thinking is if analytical thinking allows you to dissect things down to their basic components, synthetic thinking allows you to find patterns across those components. Essentially, analytical thinking will help you pinpoint the differences while systemic thinking allows you to pinpoint the similarities.

Generally, synthetic thinking is considered to be more challenging than analytical thinking because the interactions can be harder to manage, and synthetic thinking is constantly fluid whereas analytical thinking is static. While both analytical and synthetic thinking will help you find patterns and commonalities across a system or situation, synthetic thinking finds these patterns, themes, and commonalities faster and easier because it is specifically looking for them while analytical thinking is looking for differences, and they emerge as a secondary result of the analytical thinking process. It's important to understand that both analytical and synthetic thinking have their limitations and both types of thinking rely on the other.

As an example of synthetic thinking, let's once again return to the automobile industry, specifically advertising in the auto industry—an ongoing Dodge advertising campaign to sell trucks. For years, the company has identified trends in consumer information and consumer preferences to develop products and marketing strategies. These ad campaigns have synthesized several themes based on demographics, lifestyle characteristics, and psychographics. The ads promote strong and tough engines and bodies, the fact that Dodge is 'made in America' and an American company, and even feature characteristics to appeal to families because the truck can morph into a 'wagon' with some button pushing and seat rearranging. All of these parts are synthesized to persuade you to purchase a Dodge truck, no matter where you are in your life.

Analytical Thinking as a Component of Systemic Thinking

Systemic thinking is defined as an easy thinking technique for accessing systematic understanding of complicated problems and situations. Systemic thinking allows us to manage any number of elements of a situation together rather than one at a time. It gives us the ability to use systemic focus in any situation.

It's important to note that systemic thinking, systems thinking, and systematic thinking are not the same thing. These are all different types of thinking and they all have their own meaning. Systematic thinking is thinking in a methodical way whereas systems thinking is thinking about how things interact with one another. Systemic thinking is a simple technique used to identify system-wide focus.

The basic idea behind systemic thinking is that everything interacts with the system around it, that everything affects and is affected by the person or system in its immediate surroundings. When approaching things from a systemic thinking mindset, we can no longer approach problems or situations as individual parts to be managed in isolation. We now have to take the parts together and manage them cohesively. We have to manage both the elements of the situation and how the parts interact with one another.

The systemic thinking process is a four-step process that follows.

1. List as many system components as possible. This might include: problems, solutions, opportunities, needs, desired outcomes, and ideas.
2. Group common components together, and then state what each group has that makes them similar. From the above list, problems and needs might go together in the same group and desired outcomes,

ideas, and opportunities could possibly go together in the same group.

3. Find a repeating theme across group descriptions.
4. Your process should look like: Components -> Subthemes -> Common Theme. This process really kicks up your insight in any situation and you might find yourself wondering why you haven't done this sooner.

It can be incredibly difficult to recognize patterns and themes when we first encounter a challenging situation. However, with practice and as our skills develop, we should be able to recognize them more easily or at least have the skill set to manage them more adeptly. The experience of working with systemic thinking is more than just more knowledge and more skills. It's more familiarity with the patterns and themes that will give you the ability to recognize them faster and easier. It is important to note that because the ability to recognize these themes and patterns can become ingrained in our brains over time, we may start to subconsciously recognize them without cognitively being aware of them. This means we're not really comprehending what the pattern or theme means, we're just engaging in certain behaviors by rote because that's what we're used to. The idea is to purposefully find and understand the repeating patterns and themes in tough situations as this allows us to continually improve and develop mastery and better insight in the realm of systemic thinking[11].

Systemic thinking is simply a combination of analytical and synthetic thinking. The process behind systemic thinking is to list as many components as you can possibly think of and then to analyze and look for the similarities between the components. This contrasts with analytical thinking where the process is to list only a small handful of components, rank them, select only the most important one, and discard all the others.

Analyzing problems within the context of systemic theory is different from analyzing problems outside of that context. Analytical thinking insists on working with only a small number of components to keep the list you're working with manageable and to reduce the workload. Analytical thinking within the systemic thinking context depends upon listing as many different components as possible in an attempt to provide the most accurate and representative pattern. For example, in systemic thinking, the first step is analytical thinking: List as many components as possible. The second step is synthetic thinking: Find the common themes and patterns across those components.

When comparing systemic thinking and analytical thinking, systemic thinking is in direct opposition to analytical thinking. Analytical thinking breaks down systems or situations in stages whereas systemic thinking focuses on grouping similar items together. In the second stage, analytical thinking will purposefully look for the most promising component while systemic thinking looks at all of the components to find the common themes and patterns.

Developing a Goal, Problem, or Solution or GPS model can be challenging at first, but they are simple and powerful tools to give you deep insight from a neutral point of view[12]. GPS models can be completed independently or collaboratively. You will complete a GPS model in full for each component before moving on to another component and you will complete the goal wording, the problem wording, and then the solution wording. Let's say sales are down in our company that sells furniture. Our goal wording for sales staff might be to increase sales this month by 15%. The problem might be that it's Christmas season so most people are buying gifts, not furniture. The solutions might be extra financing deals or discounts for cash-paying customers. We could do this analysis for each level of employee at the furniture store: sales managers, finance managers, delivery drivers. We would look for themes and patterns that might have cost us business in the past. Maybe it cost too much for furniture delivery or the delivery was too slow. Perhaps financing rates were too high or the offer of financing too strict. All of this may become evident as we go through the GPS model.

Analytical Thinking and Critical Thinking

Critical thinking is the capacity to analyze facts, formulate and organize ideas, stand up for and defend opinions, compare and contrast items, draw conclusions, validate or invalidate arguments, and problem solve. Critical thinking is based on assumption, an assumption that there is logic involved in problem-solving and it can be figured out and reasoned through.

Clearly, critical thinking is not going to be appropriate in every situation or when trying to make every decision. Again, take the example of romance. Critical thinking and logic cannot help you determine who to date or who to marry. These are decisions that will always be made by your intuition. This is also true for other items of personal taste. I will likely never be able to explain why I love Howard Chandler Christy's World War I recruitment posters for the United States. I'm sure it's because I also like his other artwork as well and his ideal 'Christy Girl.' This isn't something I had to learn to love. It was automatic. But without a doubt there are important decisions in life that have to be thought about and mulled over such as what college to attend, whether you should even attend college at all or if trade school is a better option, or should you take that job on the other side of the country. Critical thinking is a metacognition otherwise known as thinking about thinking.

Critical thinking is important because it allows us to acknowledge our emotions, but not be controlled by them, which is especially important when it's time to make a decision. Emotions are important, especially to our memory and developing our personal tastes, but they can easily trick us too. They can mislead us into thinking we are making the right decision when we aren't. Have you ever heard that most eyewitnesses to crimes are unreliable? That's for many reasons. Sometimes it's due to outside influence, not getting as good a look at the perpetrator as the witness thought, or just the trauma of witnessing a violent crime. But it is well known amongst criminal trial attorneys and prosecutors that an eyewitness is one of the least reliable forms of evidence, and that is why circumstantial or forensic evidence to back up that witness testimony is always needed to help prove a person's guilt beyond a reasonable doubt. Luckily, critical thinking allows us to effectively handle our emotions by allowing us to sort through them and decide if they are appropriate for the current situation.

Analytical thinking is different from critical thinking in that critical thinking is more of an opinion-based type of thinking whereas analytical thinking focuses on a streamlined and focused approach to finding a solution[13]. When a person uses critical thinking skills the decision is made regardless of if the problem or situation is right or wrong. Once the critical thinker is provided with information, he analyzes the information, interprets it, and then draws conclusions using what he also knows about the world[13]. All of this comes together, and the critical thinker forms his opinion on whatever topic is being investigated. Analytical thinkers take apart pieces of information. Each item is taken step by step in an effort to develop a generalized answer to the problem being investigated. Analytical thinkers also use facts and research to support their conclusions and solutions.

Critical thinkers also have some of their own traits. According to Dr. Roy van den Brink-

Budgen, an expert in the field of critical thinking, there are four traits that are ingrained in critical thinkers: persistence, rigor, openness, and diligence[14]. In addition, Dr. Robert Ennis, Professor Emeritus at the University of Illinois, has also developed a list of skills that successful critical thinkers should be able to perform.

- Judge the credibility of sources
- Identify conclusions, reasons, and assumptions
- Judge the quality of an argument—particularly it's reasons, assumptions, and underlying evidence
- Develop and defend a position on the issue
- Ask the appropriate clarifying questions
- Plan hypotheses or experiments, and assess experimental designs
- Define terms in a way appropriate to the context
- Be open-minded
- Be well informed

- Draw warranted conclusions, but with conclusions[14]

One of the key takeaways from this list is that all of these traits or skills are teachable. Every one of them can be taught or learned. Just like analytical thinking, critical thinking isn't something you're born with it's a skill you acquire with education and practice.

A woman was near death from a special kind of cancer. There was one drug that the doctors thought might save her. It was a form of radium that a druggist in the same town had recently discovered. The drug was expensive to make, but the druggist was charging ten times what the drug cost him to produce. He paid $200 for the radium and charged $2,000 for a small dose of the drug.

The sick woman's husband, Heinz, went to everyone he knew to borrow the money, but he could only get together about $1,000 which is half *of what it cost. He told the druggist that his wife was dying and asked him to sell it cheaper or let him pay later. But the druggist said: "No, I discovered the drug and I'm going to make money from it." So Heinz got desperate and broke into the man's laboratory to steal the drug for his wife. Should Heinz have broken into the laboratory to steal the drug for his wife? Why or why not?*[15]

Some of the key questions you may want to ask yourself as you think about where you stand on this issue are…

1. The issue of price gouging, which has been in the news quite recently specifically relating to 'Big Pharma' companies that have done this. There are the infamous Martin Shkreli and Valeant Pharmaceuticals scandals.
2. How you feel about capitalism and free enterprise.
3. How much you value human life over other factors and, while you may feel bad

for Heinz and his wife, that doesn't entitle them to the drug or to steal another's property.

These are just some of the questions to consider as you approach this dilemma. Remember, critical thinkers don't worry about whether the issue is right or wrong.

Analytical Thinking and Creative Thinking

Creative thinking is relating or producing a thing or idea that was not previously related. Creative thinking comes in handy when trying to solve problems when you're panicked. Some might even be created to prevent panic. When I took my two children to Disney World, they were young enough to enjoy the amusement park, but not old enough to reliably remember their parents' names and phone numbers. I had temporary tattoos created for them with my name and phone number on them in case we got separated. Luckily, it never happened, but I got asked about where I'd gotten them from a lot of parents during our week-long vacation.

Analytical thinking is logical and leads to distinct or a low subset of answers whereas creative thinking requires your imagination and can lead to a literal fount of ideas and solutions. While each type of thinking is different, like all the other types of thinking in this chapter, creative thinking and analytical thinking are linked because they complement one another. This becomes more apparent in creative thinking when you sift through and analyze the ideas to find the few that can be put into practice. In order for analytical thinking's ideas to make progress, they must be followed by creative leaps and bounds.

Analytical thinking is convergent, meaning it pares down to a small number of distinct ideas, answers, or solutions for further analysis and utility. In comparison, creative thinking is divergent, meaning it begins with a description and then splits in many different directions to find many possible solutions.

Convergent features include: logic, uniqueness, solutions, and vertical. Divergent features include: imagination, multiple possibilities, solutions or ideas, and lateral. Using analytical thinking to approach a problem takes a deep and focused approach similar to tunnel vision to find all the necessary parts of the problem. Creative thinking is the opposite. It requires a wide-ranging approach of all possible options, including those that, on the surface, appear to be completely unrelated. I always think of creative thinking as opening the largest doors of your mind and letting everything flood in uninhibited. This is why it's thought of as lateral thinking, because the approach is wide.

Creative thinking has other benefits beyond problem-solving as well. It's actually good for you. The creative process can improve your health in a multitude of ways. One of those ways is that it's a stress reliever[16]. There are many creative activities you can do to help relieve stress in your life, and you can tailor them to your own personal interests. If you like art, you can visit museums or even visit shops looking for a new piece for your home or go for a night out at a place that helps you create a piece you can be proud of. If you've always wanted to write a book, the digital publishing age is now. You don't need an agent and a publishing house. Those days are long gone. You can be the person who knits the most beautiful and precious items when her friends have children, or you can donate items to your local NICU if knitting or crocheting soothes you. Reducing your stress levels can help you prevent heart disease, Alzheimer's disease, and depression[16]. Just have fun immersing in the creative process.

Creative thinking also increases and renews brain function. Creative tasks protect and encourage neuron growth by bolstering the growth of new neurons, which is vital in maintaining a healthy central nervous system[16]. Creative activities can also help in recovering after a major illness or injury or stress such as a breakup or divorce. This can include crafting, brain games, and even listening to music. Have you ever heard of a 'breakup' song? Songs like "So What" by P!NK or "We Are Never Ever Getting Back Together" by Taylor Swift can be played on repeat as loudly as possible as long as needed after the termination of any relationship. Learning something like cake decorating or flower arranging during a long recovery can also help you feel better as well as give you something to look forward to. Brain games in your own home can help keep your mind active so it isn't turning to Swiss cheese as you watch TV all day long.

Creative thinking also boosts your mood. No one on this Earth is 100% happy every minute of every day, and even basically happy people can experience periodic depression from time to time. I remember after my son was born my wife just cried and cried and cried, basically nonstop when she held him whilst she thought they were alone together. But as she found her footing as a new mom, she found that if she turned on some music and danced with him, he enjoyed the rocking sensation of her swaying to the beat, and she loved how he snuggled into her neck. That music and the urge to dance had a huge impact on her mood she later told me. Creative activities increase control over emotional pain and depression. It allows you a deeper understanding of yourself because you're connecting with yourself in a way that you normally couldn't or wouldn't[16].

Key Takeaways

- Analytical thinking is a cognitive process in relation to four other concepts: synthetic, systemic, critical, and creative thinking.

- Synthetic, systemic, critical, and creative thinking all work with analytical thinking (singularly or with another concept) to provide a wide range of options for approaching problems in conjunction with analytical thinking.
- Analytical thinking will help you find the differences in the components of a situation. The concepts will help you identify the similarities in the themes and patterns of the components.

Chapter 4: How to Apply Analytical Thinking in Solving Problems

The aim of this chapter is to review analytical thinking applications in a real-world setting. As analytical thinking is a problem-solving methodology and the purpose of this book is to provide you with practical applications, this chapter will cover several different areas where analytical thinking can be useful.

Errors in Reasoning

As crazy as it might sound, making a well-reasoned decision is not easy or simple. There are some common fallacies that we can become trapped in that lead us into poorly reasoned decisions. By reviewing some of these fallacies, we can reduce the chances we'll fall prey to them or at the very least we'll be able to recognize them in a timely fashion and then respond appropriately.

Synonym Errors – This error occurs when one word is replaced with another, but the words do not have the same meaning.

Non Sequitur Errors – This error occurs when the argument does not follow logic. Essentially, the conclusion is not viable with the available facts.

Red Herring Errors – This error is an unrelated reference used to distract from the argument. The term red herring derives from the use of smoked and salted fish with a strong smell to distract hounds during English fox hunts.

Unsupported Generalization Errors – This error applies specific facts to a broad generalization with no justification.

Poisoning the Well Errors – This error simply means the argument is weakened due to a criticism within the argument itself.

Cause and Effect Errors – This error sets up false choices between two choices while ignoring other possible options.

Begging the Question Errors – This error occurs with something that is believed to be true, but has not been verified, in order to support an argument.

Comparison Errors – This error looks for similarities or differences between two unrelated topics.

Questionable Authority Errors – This error occurs when using a source that is not an expert on the specific issue.

Contradiction Errors – This error states the opposite of what has been stated in the argument.

Inconsistency Errors - This is when parts of an argument are in direct opposition.

Omission Errors - This error occurs when a necessary piece of the argument has been left out.

Oversimplification Errors – This error reduces a complicated issue down to something simple.

Sampling Errors – This error relates to the data used to draw conclusions. The sampling size may have been too small or there may have been an unreliable sample group[17].

Ad Hominem Attack – This is an argument made on a person's character or situation in life instead of the argument presented.

Card-Stacking– This error is based on only using facts that show the person making the argument in a positive way.

Bandwagon Appeal – This is an attempt to conclude an argument right because 'everyone else believes it.'

Argument to the People – This error tries to win people over by appealing to the emotions of others and creating a mob mentality in the heat of the moment instead of relying on the facts[18].

Exercise: Using the error bank below, match the fallacy with the argument presented.

Ad Hominem Attack	Unsupported Generalization Error	Questionable Authority Error
Bandwagon Appeal	Argument to the People	Red Herring

		Error
Contradiction Error	Oversimplification Error	Sampling Error

1. Everybody knows that our children need the absolute best of everything if we want them to succeed in life! The best education, the best clothes, the best friends in the best social groups. All parents know that if you can't provide the best for your children you might as well not even have any.

2. Children are safest at school today than they've ever been.

3. I've just seen four doctors in this hospital and all of them were men with brown hair. I guess this hospital only hires men with brown hair to be doctors.

4. Psychic Cindy will be in charge of helping ensure all safety and structural guidelines are followed at our next build project, so we don't have another disaster like we did last time.

5. You can tell just by the length of Rachel's skirt that she'll sleep with anyone. She's not a good person at all. Just look at how short her skirt is.
6. Barbara promised to come to the party today. Barbara doesn't see why she should go to the party today.
7. Our country is in shambles. We are broke, being lied to on a daily basis by our president, we are divided by our leaders, and every day it gets worse. We need to vote these

crappy politicians out of office and elect new ones who will put the average American's interests first.

8. I just read a research study that studied the prevalence of skin cancer in fair-skinned redheads. I was really interested until I saw the researchers only had 5 test subjects.
9. I hate football. One team throws the ball that way. The other team throws the ball the other way and sometimes they kick it.

Analogies

An analogy is a comparison between two things or systems of things that point out the ways the items are similar. Analogical reasoning is any type of reasoning that relies on this type of comparison and then further argues that other similarities between the two systems must exist as well. Analogies draw a comparison between the shared characteristics of the two items. This, in turn, requires us to use analytical thinking skills to analyze and evaluate forms, usages, structure, and relationships. Understanding relationships and being able to create corresponding relationships are priceless skills you'll use your entire life.

You may be asking why learning analogies are so important. It's because analogies are the foundation of thinking. We use analogies on a daily basis in both our written and spoken communication with others, and they are absolutely crucial to the languages of the arts and sciences. Analogies are also routinely used to assess verbal skills, creative thinking, and analytical thinking. In addition, analogies can help you solve difficult problems when you are short on time. This is known as a heuristic role. Analogies can be used in endless settings, with the intent to develop insight and create possible solutions.

Examples of analogies

a) One way analogies can be used is to persuade people to take an idea or problem seriously. An example is Charles Darwin using an analogy between artificial and natural selection to support his argument for natural selection.

   Why may I not invent hypothesis of Natural Selection (which from analogy of domestic productions, and from what we know of the struggle of existence and of the variability of organic beings, is, in some very slight degree, in itself probable) and try whether this hypothesis of Natural Selection does not explain (as I think it does) a large number of facts...[19]

   In this example, Darwin uses an analogy to argue that his hypothesis is plausible to at least some degree and therefore warrants further research in this area.

b) It's also possible for analogies to provide poor support for the conclusion they're meant to bolster. In Thomas Reid's argument for life on other planets in 1785, Reid based his argument on the similarities of other planets to Earth, such as they all orbited around and were illuminated by the sun, some had moons, and all of them turned on an axis. Reid concluded that it wasn't 'unreasonable to think' that life existed on other planets.

c) Another example is rectangles and boxes. Let's imagine that you've come to the conclusion that of all the fixed four-sided shapes, you've determined that the square has the maximum area. Based on this conclusion, you then surmise that of all four-sided shapes, the cube has the most volume.

d) In the mid-1930s Schaumann, a pharmacologist, was testing synthetic compounds for their anti-spasmodic effects. He noticed that one of these compounds, Demerol, had a similar effect to morphine and he theorized that the drug might also have the same painkilling effects as morphine. After extensive testing, this was determined to be the case, and Demerol is used worldwide as a safe and effective painkiller[20].

The Structure of Analogical Arguments

An analogical argument takes the form below.

S is similar to T in certain known ways

S has some additional feature Q.

Therefore, T also has the feature Q or some feature Q*, which is similar to Q

Lines one and two are the premises and line three is the conclusion drawn from the information in lines one and two. The logic and argument form uses inductive reasoning as the conclusion may or may not follow the premises. S and T are known as the source domain and the target domain. A domain is a set of objects, characteristics, relationships, and functions all together with a group of accepted statements about those same areas. Technically, an analogy between S and T is a one-to-one parallel between the objects, characteristics, functions, and relations in S as well as those in T. Additionally, not all of the items in S need to be compared to T. You would generally look at analogies for the most important similarities or differences. For example, if we revisit Reid's argument from earlier, his analogy using the structure above would look like this…

↑ **Earth (*S*)** **Mars (*T*)**

vertical Known similarities:

	orbits the sun		orbits the sun
	has a moon	←	has moons
	revolves on axis	**horizontal** →	revolves on axis
	subject to gravity		subject to gravity
	Inferred similarity:		
↓	supports life	⇒	*may* support life[21]

According to Hesse, horizontal relations in an analogy are the relations of similarity and difference in the mapping between domains, which you can see in our example above[21]. The vertical relations are those between objects, relationships, and characteristics in each domain. In our example, there is a horizontal relation between Earth having a moon and Mars having a moon whereas there is a lesser vertical relation to Earth supporting life and Mars supporting life.[21]

Skill Steps to Analyze Analogies

- Analyze and determine the relationship between the two objects that are being linked. Be sure you include all the possible answers to the analogy as this can provide useful information when determining the correct conclusion.
- If possible, put the relationships into categories or groups.
- Develop possible answers and look for solutions among the solutions being offered. Don't forget that you may still find better options.
- Investigate the relationships between the pair of items that are linked, considering all possible combinations.

- Determine the best solution by selecting which combination fits best between the linked objects.

The Metacognition Step

- Think about the process you used when you tackled this problem. What thinking skills did you use and were they effective?
- What about your thinking process worked?
- What about your process didn't work?
- Next time, what will you do differently to get a better or faster result and improve upon your analytical thinking?

Guidelines to Interpret Analogies

Interpreting analogies is almost a science unto itself. It has rules and guidelines written in a textbook style by the field's logicians and scientists with the aim to provide direction for evaluating analogical arguments. Here are some of the most important ones.

(G1) The more similarities between two domains, the stronger the analogy.

(G2) The more differences, the weaker the analogy.

(G3) The greater the extent of our ignorance of the two domains, the weaker the analogy.

(G4) The weaker the conclusion, the more plausible the analogy.

(G5) Analogies involving causal relations are more plausible than those not involving causal relations.

(G6) Structural analogies are stronger than those based on superficial similarities.

(G7) The relevance of the similarities and differences to the conclusion (i.e., to the hypothetical analogy) must be taken into account.

(G8) Multiple analogies supporting the same conclusion make the argument stronger[22].

While these guidelines are helpful, it's important to remember they aren't without their flaws. One of their biggest handicaps is that they are simply too vague to provide any real or helpful guidance or insight. How do we know what to count as a similarity or difference in terms of application to G1 or G2? What's particularly special about the structural and causal analogies mentioned in G5 and G6, and why do we need to give them extra attention? And finally, how do we know which analogies that have already been removed from the argument warrant a second glance? The guidelines provided do little to nothing to educate readers on these topics, which for a novice, is incredibly important for building analytical thinking skills.

Simple Exercises[23]

Marshal : prisoner ::

Principal : ____________

a)teacher

b)president

c)doctrine

d)student

Denim : cotton ::

________ : flax

a)sheep

b)uniform

c)linen

d)sweater

Segregate : unify ::

Repair : __________

a)approach

b)push

c)damage

d)outwit

Monkey : primate
_________ : marsupial ::

a)opossum
b)ape
c)honeybee
d)moose

Monarch : ________ ::
King : cobra

a)queen
b)butterfly

c)royal

d)venom

Now let's check your answers. The answers are: d – student, c- linen, c – damage, a-opossum, and b-butterfly. Did you get them all right? Let's go over them one by one. In the first analogy, a marshal is in charge of a prisoner and a principal is a person who is in charge of a student. In the second analogy, cotton is used to make denim whereas flax is used to make linen. One of my favorite analogy solving tricks is using one pair of words to make a sentence, as it will often underline the relationship between the pair, and then using that same sentence to see which answer works best. In the third analogy, segregation is the opposite of unify and repair is the opposite of damage. Next, we can see that monkey is a type of primate, and opossum is a type of marsupial. In the final exercise, king is a type of cobra just as monarch is a type of butterfly.

Now let's try some more challenging exercises[24].

Castigate : criticize ::

a)dishevel : destroy

b)brutalize : attack

c)legalize : filibuster

d)predict : forecast

e)kill : massacre

Virulent : poisonous ::

a)autonomous : free

b)disparate : different

c)botanical : flowery

d)dormant : tired

e)unique : eccentric

Invective : blame ::

a)homage : copy

b)grandeur : admire

c)duplicity : increase

d)depravity : corrupt

e)masonry : construct

These exercises were much harder because we needed to find the entire second pairing of the analogy. Let’s see how you did with these. The answers here are b – brutalize: attack, b – disparate: different, and e – masonry: construct. To explain these answers, let’s look at the first one. Castigate is an extreme form of criticizing. If you consider all your options, you’ll realize that brutalize is an extreme form of attack, making the answer b correct. The second example is very similar to the first. Virulent means something severe or especially harmful, so the first part of the analogy relates to being extremely poisonous. Of the choices available, disparate relates back to being different in a severe or extreme form, making it the correct choice for this exercise. In the final exercise, invective is abusive and hateful language used to blame someone. This leaves the only viable choice as masonry because masonry is used to construct.

Analyzing Trends and Patterns

Trends are the general direction of a unit, such as a dollar or shares of stock, over a period of time. The three directions trends move in are up, down, and horizontally. Trends can be found in the short, mid, and long terms. Analysts will use a trend line in an attempt to identify and define trends. Patterns are a dataset that follows a recognizable form, which analysts look for in current data. We see patterns as datasets that reappear in recognizable ways[24]. We may see patterns mark the beginning of an upward or downward trend or possibly even a new trend. Trends and patterns can be seen in numerous areas such as politics, economics, finances, linguistics, and even statistics.

Statistical Analysis

Humans generate an enormous amount of data on a day-to-day basis. In this day and age of technology, everything you do generates data. Each time you post on social media, send a text, make a purchase online or with your credit or debit card you are generating data. There are even apps that have been built that are running in the background that are collecting data on how you use your smartphone and sending it back to the company that developed the app and most consumers aren't even aware of it. It's said humanity now creates as much data every day as we did from the dawn of civilization up until the year 2000. At the rate we are relying on our smartphones and other gadgets our 'data footprint' won't be decreasing any time soon.

So how are we to make sense of this massive amount of data pouring into our laps on a daily basis? Let's say you own a company that makes warm and comfortable pajamas for the entire family and you sell your pajamas to retail locations all over the country. One of your sales staff tells you that the company made preorders to 500 retail locations in preparation for the coming 'Black Friday' shopping day after Thanksgiving. If you want to look over the preorders to see where your pajamas are having the most success, you don't want to spend the next six months eyeball deep in paper going over 500 different sales reports. You want to see a summary of the figures. You may want to know the total of all the preorders and their average value, and later, you'll want to compare this information with actual orders fulfilled to so if any orders were canceled. You may also want to know what just the ten largest sales were or note the top selling items your company is on track to sell for the holiday season so you can be sure to have plenty on hand and ready to ship so you don't lose business to competitors. You'll be able to find all of this information quickly and efficiently if you use statistical analysis.

Statistics is a discipline within mathematics that involves the collection, analysis, interpretation, and presentation of data for decision-making purposes. Many organizations employ statistical analysis as a way to try and predict future trends.

Descriptive Statistics

Descriptive statistics delivers on its name and describes existing data using measures such as sum, mean, median, and others. In the example of the sales reports, we can calculate mean (average) and sum (total earned) to see how the business is doing in terms of the overall income we're bringing in and to see on average, how much retail locations are ordering. Black Friday gets its name because it's the day most clothing retailers go from operating at a negative profit (in the red) to finally making a profit for the year (in the black). So this time of year is when we can expect to see our business increase significantly. We can also calculate the median (what orders were at the exact center of all the orders), and in this process look at the smallest orders and the largest orders. If the smaller orders were exceptionally small, we may want to look at metrics on the overall performance of pajama sales nationally. Pajamas and related seasonal clothing are very popular holiday gifts as well as necessary personal purchases as the weather changes. If statistics show our sales patterns are significantly different from previous years, we'll want to look for other reasons as to why our sales are faltering.

The benefit of descriptive statistics is that it allows you to summarize large amounts of data quickly and efficiently. You probably already use some of these in your day-to-day life. When you say something like, "On average I spend $650 per month at the grocery store," you're using descriptive statistics by doing some quick statistical analysis in your head. You use them when you calculate your MPG, if you're a server in a restaurant you may count out your tips for the night to see how much you earned per hour. All of that is routinely using descriptive statistics by the average person, not someone with a degree in statistics.

Inferential Statistics

Inferential statistics tries to surmise something about the data such as patterns and relationships. This type of statistics often requires statistical testing. Let's return to our original example. . Let's say we want to increase the sales at our pajama company and since we sell our pajamas to retail locations all over the country we want to run Facebook ads to increase the demand for our product, and since Black Friday is the official start of the holiday shopping season, that is when we want to launch our new campaign.

Running Facebook ad during the heaviest retail-shopping period of the year is expensive, so we've decided to start small and work our way up. We've decided to start with a small daily budget for these ads and target all of our key words independently in an effort to determine which ones are effective and which ones have minimal impact. Facebook ads allow us to target specific age groups and genders and we decide to target women between the ages of 25 and 75, as we know from industry standards this is the primary consumer of our product and the buzzwords for ads will be targeted at specific age groups within this range. After giving specific keywords a reasonable amount of time to make an impact on our sales figures, we can calculate how much each targeted ad increases sales during the ad campaign and select the top performing ads for additional statistical analysis. Why do we need further analysis if we know which buzzwords on Facebook ads performed the best? Because in reality sales can and do fluctuate for any reason and that's just par for the course as a clothing manufacturer. Maybe sales had increased because of particularly good sales prices offered at one of our retailers or because our competitor's products were at a higher price point. How can we pinpoint if any increase is due to the television ads or some other reason?

This is a job for inferential statistics! We're going to run a statistical test to help us determine how confident we can be in the effectiveness of our Facebook ads. We're going to need to divide our numbers into two groups: one for an ad that was geared toward women in the 50 plus age groups as our industry research shows that grandmothers and other extended family members often give pajamas to younger generations as gifts during the holiday season, and one group that looks our sales numbers for this exact same period of time last year, when we were not advertising on Facebook, but were paying for features in retailers' print ads. Then you conduct a test to compare the groups. For our purposes, we're not going to worry about the test as there are many types of statistical tests you can run depending on the many different needs of the researcher.

After the test is concluded, our results would look something like this: I am almost sure the Facebook ad campaign targeted towards the 50 plus age group led to an increase in sales by ten percent or more during the weeks we ran the targeted ad campaign. Why aren't we absolutely sure? In statistics you can never be absolutely sure because there will never be enough data to give you an absolute answer. However, you can take your data and decide whether or not the pajama company's Facebook ads were or were not a good investment to increase sales.

And last, but not least, you can also have inconclusive results. If the sales had only increase may be one to three percent, you may not have enough confidence to make a decision as to whether the campaign was successful or not. While that's not terribly helpful, it's better to find that out during a controlled experimental campaign rather than after a costly ad cycle that did not reap any benefits for the company.

Predictive Statistics

Predictive statistics attempts to predict future information based on the current data we have access to. Many of us use one type of predictive statistics on a daily basis: the stock market. Stock brokers evaluate historical data and world-wide events, such as political conflicts and oil production in the case of oil and energy stocks, to make a prediction on whether particular stocks will rise or fall and provide a good and reliable investment over time. It's not always 100% accurate, and there have certainly been some exceptions, such as the stock market crash in 1929 that lead to The Great Depression, or stocks tanking after September 11, 2001, and the impact the financial scandals and government bailouts in 2008, but overall you should be able to make an informed and educated decision on which stocks based on a long term and steady growth plan.

Now let's return to our running example of our pajama manufacturer. One of our competitors is coming out with some snazzy new designs and styles and they look like they will probably sell well. We want to stay on top, so we've lined up some new items as well such as soft, extra-cushioned slippers, matching robes, and our men's line has a more rugged look we think guys will enjoy. One of our sticking points is trying to determine what colors to use. We don't want to get stuck with mass back stock of camo robes or fuchsia slippers that we have to sell at steep discounts. How do we decide which colors or styles will appeal to our consumers?

To get the information we need to determine what colors and patterns are most likely to be successful, we can pull up sales figures from past product launches and see what items were hot and which items were not. The items may have been different, but we were targeting the same type of consumer. If we analyze our data—Which style sold best? Which colors were most popular?—we can develop a statistical model to help us plan out what items of our new line to put into production and what colors and fabrics our customers like best. Is this statistical model going to be 100% accurate? No, it isn't, but using this type of model will hopefully be all you need to give you that boost over your competitors when the new lines are launched, and it's a much better strategy than launching a new product blind with little or no strategy.

Statistical Software

Statistical software is a necessary tool for statistical analysis. Spreadsheet software typically includes a number of standard statistical functions, but more advanced statistical analysis will require more specialized software programs. The good news is that some of these programs can be purchased at a low cost for a specific period of time if you don't anticipate needing the software for an extended period of time. There are also some newer software programs that have come out in the wake of some of the larger more expensive software companies to offer a great product at a much more reasonable cost if you're willing to do a little digging on the internet.

One of the first steps to analyzing your data is organizing and managing it. Consequently, many statistical software programs also function as a database management system (DBMS)[25]. It should be noted that if your dataset is large and complicated that you may prefer, and it may be wise, to manage your data as an actual DBMS and only transfer the parts as they are needed to the software for analysis. Statistical software typically includes functions that allow you to run descriptive, inferential, and predictive analysis, and even allows you to create graphs and other charts for the purpose of explaining your data[25].

Statistical Analysis

By definition, statistical analysis is the study of analyzing large amounts of data to explore the underlying patterns, trends, and insights hidden with them[26]. Statistical analysis comes in two formats: descriptive and inferential. If you think back to the types of statistics we discussed in our pajama manufacturer example, it may be easy to get confused with descriptive or inferential statistics vs. descriptive or inferential statistical analysis. Don't worry, the differences are easy to understand. Here we are talking about an analysis of data we've collected and run calculations on. Descriptive analysis is used to summarize data that is currently available, and inferential analysis is used to surmise insights from data that isn't currently available. It used to make judgments and glean insights from the data[26].

Examples of Statistical Analysis

Let's say Jake owns a small chicken farm and he' specializes in supplying farm fresh, free range, organic eggs to his local community. He owns 50 hens and he's asked his employees to monitor the egg output by each of the hens for one month so he can determine if additional hens need to be purchased and current hens need to be retired to the pasture to enjoy the rest of their natural life.

Measures of central tendency are a type of descriptive analysis and are used to represent the typical scenario the data illustrates. Median, mean, and mode are the three most frequently used measures of central tendency.

Median is calculated by arranging the data points in ascending order and then taking the middle number. In the case of two middle numbers, their average is taken.

Mode is the most frequently occurring data point in the dataset. In our example, Jake finds out that ten tutors were given the score of 100, and this was the most frequently occurring score. In such a case, 100 would be the mode.

Measures of dispersion are descriptive analysis and are used to explain how far apart the data points are spread. The most commonly used measure is standard deviation.

Standard deviation is a measure of how far data lies from the mean. Calculating standard deviation starts by taking the difference of each data point from the mean of the data, squaring them, then adding them up. Then, square the sum by the number of data points and then taking the square root, as shown here.

$$Standard\ deviation = \sqrt{\frac{\sum(x - \bar{x})^2}{n}}$$

One good benefit of statistical software is that it can calculate the standard deviation for you. With big data sets, statisticians have few other options. Could you imagine the time it would take to hand calculate your data and then check your work?

Standard deviation is incredibly useful in a normal distribution. A normal distribution is a variable that is distributed evenly about a mean. It is usually bell-shaped and symmetrical about the mean. Below you can see an example.

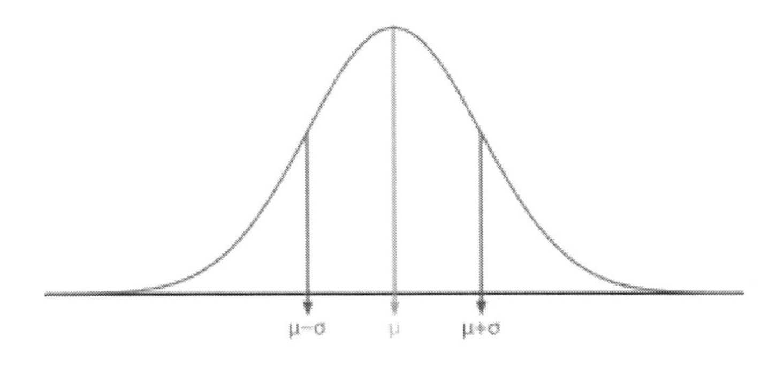

In a normal distribution, approximately two-thirds of the data points lie between one standard deviation above and below the mean. In our example, if the eggs Jake's hens are producing are normally distributed, the mean score is 60 and the standard deviation is 20, then two-thirds of the students have scored between 40 and 80.[26]

Tests of difference are inferential statistical analysis and help determine whether the difference between various groups in the data sample has occurred randomly or if it is due to another variable.

Two regularly used tests for this are the t-test and the t-test calculator.

A t-test determines whether the difference between the averages of two groups in a dataset is statistically significant, which means it is unlikely to be due to random chance. A t-test calculates a ratio called a t-value to determine whether the difference is large enough to be significant.

For example, on our chicken farm, hens younger than three years laid 65 eggs on average in one month while hens that were older than three years of age laid an average of 55 eggs in one month. It could be because younger hens produce more eggs than older hens or a chance occurrence. To test this out, Jake would have to run a t-test on the egg production numbers of the two groups. If the t-value turns out to be significant, he can conclude that younger hens lay more eggs than older hens.

A *t*-test asks if the difference between the averages of two samples is significant enough to say that some other characteristic could have caused it.[27]
To conduct a t-test using an online calculator, complete the steps in the list below.

This information was gathered by Dr. Patrick Biddix, an assistant professor in Higher Education and Research Methodology at the University of Missouri – St. Louis. Using the t-test calculator at the QuickCalcs website (link below) allows you to quickly determine if your dataset is significant and if there is a relationship between the two variables. This is great because, like statistical software, a computer program does the calculations for you, and you don't have to personally calculate and recalculate your numbers and figures. This is a tool the everyday person can use without having to be a statistician or someone who has an in depth knowledge in research methodology. All you need is a little understanding of what is being calculated and how what that figure means, which this book has provided

"Step 1. Compose the Research Question.

Step 2. Compose a Null and an Alternative Hypothesis.

Step 3. Obtain two random samples of at least 30, preferably 50, from each group.

Step 4. Conduct a t-test:

- Go to http://www.graphpad.com/quickcalcs/ttest1.cfm
- For #1, check "Enter mean, SD and N."
- For #2, label your groups and enter data. You will need to have mean and SD. N is group size.
- For #3, check "Unpaired t-test."
- For #4, click "Calculate now."

Step 5. Interpret the results (see below).

Step 6. Report results in text or table format (see below).

- Get p from "P value and statistical significance." Note this is the actual value.
- Get the confidence interval from "Confidence interval."
- Get the t and df values from "Intermediate values used in calculations."
- Get mean and SD from "Review your data."[27]

Interpreting a t-test

t	tells you a *t*-test was used.
(98)	tells you the **degrees of freedom** (the sample – # of tests performed).
3.09	is the "*t* statistic" – the result of the calculation.
$p \leq .05$	is the probability of getting the observed score from the sample groups. **This is the most important part of this output to you.**
If this sign	**It means all these things**
$p \geq .05$	likely to be a result of chance (same as saying A = B)
	difference is not significant
	null is correct
	"fail to reject the null"
	There is no relationship between A and B.
If this	**It means all these things**

sign	
$p \leq .05$	not likely to be a result of chance (same as saying $A \neq B$)
	difference is significant
	null is incorrect
	"reject the null"
	There is a relationship between A and B.[27]

Key Takeaways

- Beware of fallacies in reasoning and make sure you're not getting trapped in these errors as you're going through your own thinking process.
- Analogies and analogical thinking are foundations of thinking. We use them daily in many ways. Practice them for fun and to keep your mind sharp and your thinking process strong.
- Statistics is the tool we use to acquire information before we make decisions.

There are many different types of statistics and statistical analysis and even software options to help us analyze our data. The information we get from statistics can help us make informed and educated decisions rather than foolish ones that cost us time and money.

Chapter 5: System Analysis – Policy Analysis

Systems analysis entails gathering and evaluating facts, diagnosing problems, and the breakdown of a system into its individual elements.

The goal of systems analysis is to look at a system or its elements and then find its purpose or function. It's a problem-solving strategy aimed at improving the system. If an analysis is successful, it can help the examiner identify ways to improve the elements of the system so they can work well together and accomplish their goal. Analysis shows what the system ought to do.

There are two distinct approaches to what systems analysis is. One is "the process of studying a procedure or business in order to identify its goals and purposes and create systems and procedures that will achieve them in an efficient way." (Merriam-Webster)

In this approach, systems analysis is close to requirements analysis or operations research.

The other approach describes system analysis as a problem-solving technique. When we analyze a system, we break it down into its elements to be able to study how well those elements interact to accomplish their purpose.[28]

As a quick summary, a system by definition has three main parts:

- elements;
- interconnections; and
- a purpose or function.[29]

System analysts use *methodology* with the systems included in the analysis, and can conclude an overall picture as a result. Methodology can be defined as the systematic, theoretical analysis of the methods applied to a field of study. It incorporates the theoretical analysis of the body of methods on one hand and the principles associated with a branch of knowledge on the other hand. It contains concepts like paradigm, theoretical model, phases, and quantitative or qualitative techniques.[30]

It's important to remember that methodology and method are not synonyms. Unlike a method, a methodology's purpose is not to give solutions to a problem. A methodology gives the theoretical framework to understand which method or best practices should and could be used in a given case. Using these terms interchangeably can lead to confusion and misunderstandings. Furthermore, it can undermine a good analysis, which needs to focus on research and viable routes to improvement.

Merriam-Webster defined methodology as follows:

- "the analysis of the principles of methods, rules, and postulates employed by a discipline"; and
- "the systematic study of methods that are, can be, or have been applied within a discipline."[31]

The methods we identify in the methodology will provide the means of data collection, or how to calculate something in the hope of a specific result. Methodology, when used correctly, gives us a constructive generic framework. Within this framework we can break down the methods indicated into sub-processes—we can combine, manipulate, and study them.

We can use systems analysis in any field where something is developed. Let's have a look at a few examples of fields where systems analysis is applied:

Policy analysis

Robert McNamara, the former Secretary of Defense of the United States, instituted the discipline of policy analysis. He originated it from the application of systems analysis.[32]

Policy analysis is a procedure utilized in public organizations to empower government workers, activists, and others to inspect and assess the accessible choices for executing the objectives of laws.

The procedure is additionally utilized in the administration of big organizations with a wide range of policies. It has been characterized as a way toward figuring out which of the different policies will accomplish the given objectives, all while considering the relations between those policies and objectives.[32]

The methodology of policy analysis consists of both qualitative and quantitative strategies. The qualitative research is composed by contextual analyses, case studies, and interviews. Quantitative research incorporates surveys, measurable examination (data analysis), and model designing.

A typical practice is to clearly define the issue and assessment criteria; distinguish and assess options; and suggest a specific policy based on the results. Advancement of the best plans are the result of cautious analysis of policies by a priori and a posteriori assessments.[32]

Dimensions for analyzing policies[34]

We distinguish six dimensions in policy analysis. These are sorted as the impacts and executions of the policy over a certain timeframe. We also assess the so-called durability of the policy, which means the potential in the substance of the policy to create noticeable and effective change or results over time.

The six dimensions can be separated in two groups: effects and implementation. Policy makers usually ask these questions and design or adjust the policy according to the answers.

Effects

“Effectiveness:
What effects does the policy have on the targeted problem?

Unintended effects:
What are the unintended effects of this policy?

Equity:

What are the effects of this policy on different population groups?

Implementation

Cost:

What is the financial cost of this policy?

Feasibility:

Is the policy technically feasible?

Acceptability:

Do the relevant policy stakeholders view the policy as acceptable?"[33]

The three dimensions attached to the 'effects' category can represent some confinements because of information collection. The 'implementation' dimensions impact acceptability. The level of acceptability depends on the conceivable definitions of actors engaged in feasibility. In the event that the feasibility dimension is undermined, it will put execution and implementation in danger. This can involve extra expenses. Implementation dimensions impact a policy's capacity to create results or effects.

In other words, the 'effects' dimensions are responsible for the design of the policy and the 'implementation' dimensions will have the real say in whether or not a policy will or can be actually enacted.

The six dimensions of policy analysis are broadly usable analytical tools. You can use them in any new life rule or goal you have to make.

For example, let's say you're trying to bargain more home office days from your boss because you feel that you're getting too distracted by coworkers in the office. Let's see how that request would play out within the framework of the six dimensions:

Effects

“**Effectiveness:**

What effects does the request have on the targeted problem (distraction by coworkers)?

If your wish is granted, you would be able to work in solitude more days of the week. Your productivity may rise, you could choose to work in your most productive hours, rest in conditions that might not be fitting in an office, and manage your time and energy better overall.

Unintended effects:

What are the unintended effects of this request?

You might become distracted by other things such as television and chores. You might procrastinate knowing no one will look at you with crossed eyes. Also, your coworkers might get envious of you having more home office days and might treat you with distance—or they'd also demand more home office days.

Equity:

What are the effects of this request on different coworkers?

We answered this question partially before: the reaction of your coworkers might not be positive. They may or may not ask for more home office days, but the mood in the office could get bitter, and people might get demotivated by your "special treatment." They could also get motivated, thinking if they work hard they could earn some special treatment, too. Both of these attitudes could surface at the same time. The ratio of them would depend on the character makeup of each coworker.

Implementation

Cost:

What is the financial cost of this request?

In theory, there would be no financial cost to this particular request. However, if you end up not working (enough) on your home office days it might be a net loss for your company as they pay you the same amount for less work.

Feasibility:
Is the request technically feasible?

Yes, if you can complete your task at home and you are not needed in the office for anything else, this request is feasible.

Acceptability:
Do the relevant request stakeholders view the request as acceptable?

In this case the stakeholders are you and your boss. On your end, this request surely is acceptable—what's more, desirable. If you knew that your request wasn't feasible, you probably wouldn't even ask for it. So it all comes down to your boss: will he or she find your request acceptable?

You can brainstorm about it in advance. Is there enough workforce at the company, do you often get last minute assignments that require your presence in the company, do you have any special skills that would make your physical presence essential, and so on. The best way to find out what your boss thinks is to ask them.

Going through the phases of this analysis might highlight some aspects of your request that you didn't think about when you enthusiastically came up with it, such as the reaction of coworkers, your covert role in the company's blood flow, and getting even more distracted at home. You can decide whether the threats and costs are worth the benefits in the first place. If yes, consider how to present your case as comprehensively as possible to achieve the desired outcome.

The Five-E method

Let's take a look at a policy analysis model called the Five-E method. The five Es stand for effectiveness, efficiency, ethical considerations, evaluations of alternatives, and establishment of recommendations for positive change.

Effectiveness

When we talk about policy effectiveness we should analyze how well it works or if it would work at all.

Efficiency

To see how efficient a policy is, we need to assess how much work it will involve. Are there noteworthy expenses related to this policy, and would we say they are justified, despite all the trouble?

Ethical considerations

Is the policy ethically and morally solid? Are there any unintended consequences? Could there be?

Evaluations of alternatives

How fitting is the policy in comparison to other alternatives? Have all the significant different policies been considered?

Establishment of recommendations for positive change

What can actually be implemented? Is it better to correct, substitute, expel, or include a policy?[34]

This model of analysis involves similar elements as the previous one—the six dimensions of policy analysis—but here we introduce an important element, namely the concept of alternatives. Using the previous example, bargaining for some more home office days at your workplace, let's see what you should consider using these five steps:

Effectiveness

When we talk about policy effectiveness we should analyze how well it works or if it would work at all.

This is basically the same step that you used in the six-dimension analysis, namely, if your request is granted, you would be able to work in solitude more days of the week. Your productivity may rise, you could choose to work in your most productive hours, rest in conditions that might not be fitting in an office, and manage your time and energy better overall.

Efficiency

To see how efficient a policy is, we need to assess how much work it will involve. Are there noteworthy expenses related to this policy, and would we say they are justified, despite all the trouble?

This is a different perspective than in the previous analysis. Would the request by itself generate less workload? Unlikely. You would still need to finish your tasks. But your productivity could improve, thus you'd win some time to do even more work or to have some 'me' time. Also, the fact that you don't have to commute there and back to your workplace makes life easier and wins you even more time. Financially, the answer here is the same as in the previous analysis: it wouldn't cost your company more to have you in the home office unless you become less productive.

Ethical considerations

Is the policy ethically and morally solid? Are there any unintended consequences? Could there be?

The unintended consequences are the same here as before, your coworkers might get annoyed with you and even envious. The question of morality and ethicality is an interesting one which you should consider before filing a request for more home office days.

For example, does anyone else in your position have home office days? Could it be a burden the company or your coworkers should bear? If there are more people competing for more home office days, is your request more warranted than that of your coworkers, like a mom who has a two-year-old at home or a sick parent or spouse who needs constant attention?

Evaluations of alternatives

How fitting is the request in comparison to other alternatives? Have all the significant different alternatives been considered?

This is a good analysis point for brainstorming to have a plan B up your sleeve. For example, requesting a half-day at home or more flexible working hours instead of full-day leaves. If you can't have it all, at least push for some improvement in your life.

Establishment of recommendations for positive change

What can actually be implemented? Is it better to correct, substitute, expel, or include the request?

After you present your request and your alternatives to your boss, you can talk about what is feasible and what isn't. Maybe you can get to a midway understanding. Let's say you requested one more home office day each week. Maybe you get two extra home office days per month, plus two days with flexible hours. Or you get the four home office days but in exchange your daily quota will be raised by 10% on the home office days. You can bargain to find a mutually satisfying middle ground.

Key Takeaways:

Policies are frameworks that can enhance general prosperity in a system or organization. Authoritative bodies and lobbyists make the policy analysis with the aim of bringing an evaluative result. Policy analysis needs to look in-depth at social issues that need attendance. These are the steps, which policy analysts follow:

1. "Defining the problem assessed by the policy;
2. Assessing policy objectives and its target populations;
3. Studying effects of the policy;

4. Policy implications: distribution of resources, changes in services rights and statuses, tangible benefits; and
5. Alternative policies: surveying existing and possible policy models that could have addressed the problem better or parts of it which could make it effective."[35]

In Closing…

Analytical thinking is a priceless skill. It's a commodity that everyone wants, not everyone has, but if everyone took the time, everyone could learn. It's important to remember that above all else analytical thinking isn't a genetic trait, but a skill to be learned just like any other. Just as you learned to ride a bike as a child and drive a car as a teen, you can also learn to think analytically. It's a truly important thinking process because once you have taken the time and invested in this process, it will repay you over and over. Analytical thinking will help you think logically and make clear-cut analytical decisions in your life. You can take this skill with you to purchase a car, to negotiate for a higher salary by making a logical and concise argument about why you've earned that salary, and even researching expensive or special purchases before you make them, so you know you're making a purchase you'll be happy with.

Analytical thinking skills can assist you in making positive money management decisions so you're prepared for retirement and don't have to work longer than you anticipated. Money management skills are also vital for those nowhere near retirement. Younger people need to have an emergency savings account, but then a larger one in case you find yourself without work for three to six months. Most people simply glide through life thinking these things will never happen to them. I hope they don't, but if they do, thinking analytically will tell you that you're no more immune to these issues than any other person. But using your analytical thinking skills can help protect you because you will make wise decisions that will help you be prepared.

I encourage you to practice the skills we've learned in this book, all of them. Some of them are fun, such as the practice analogies and incorporating creative thinking and activities into our analytical thinking. Some of them are a little more bookish and will require a little more intensity. But don't worry. All of them are designed to help you become a better analytical thinker. Mix up the activities and enjoy yourself. Try to do something that uses these skills every day until analytical thinking is second nature and you don't even need to think about it anymore, because that's the goal, to make analytical thinking an ingrained part of your thinking process.

And last, but certainly not least, thank you for selecting this book. I hope it was both informational and enjoyable. I hope you learned something new if not many new things. My goal is to help you enhance your life and live the best possible life you can. What could be better than that?

References:

1. Ludden, David. PhD. Are You an Intuitive or Analytical Thinker? Psychology Today. 2016. https://www.psychologytoday.com/us/blog/talking-apes/201602/are-you-intuitive-or-analytical-thinker

2. Tang, Rong; Sae-Lim, Watinee (28 July 2016). "Data science programs in U.S. higher education: An exploratory content analysis of program description, curriculum structure, and course focus". *Education for Information*. **32** (3): 269–290. doi:10.3233/EFI-160977.

3. Terrel, Shannon. 5 Of The Best-Performing Analytical Skills For Your

Resume. Mind Valley. 2018. https://blog.mindvalley.com/analytical-skills/

4. Kagatikar, Mukund. Top 5 Benefits of Analytical skills. LinkedIn. 2015.https://www.linkedin.com/pulse/top-5-benefits-analytical-skills-mukund-kagatikar-csm-scrum-mentor

5. Parselle, Charles B. Analytical/Intuitive Thinking. Mediate. 2005. https://www.mediate.com/articles/parselle6.cfm

6. Al-Shammari M. (1995) Teaching Problem-Solving And Analytical Thinking Skills: With A Special Reference To Modelling In Management Science. In: Gijselaers W.H., Tempelaar D.T., Keizer P.K., Blommaert J.M., Bernard E.M., Kasper H. (eds) Educational Innovation in

Economics and Business Administration. Educational Innovation in Economics and Business, vol 1. Springer, Dordrecht.

7. Doyle, Alison. Analytical Skills Definition, List, and Examples. The Balance Careers. 2018. https://www.thebalancecareers.com/analytical-skills-list-2063729

8. Rahul. How to Build Analytical Skill. 2018. https://zigya.wordpress.com/2018/06/07/how-to-build-analytical-skill/

9. Satyendra. Analytical Thinking Skills For Problem Solving. Ispat Guru. 2016. http://ispatguru.com/analytical-thinking-skills-for-problem-solving/

10. Amer, Ayman. Analytical Thinking. Center of Advancement of Postgraduate

Studies and Research in Engineering Sciences. Faculty of Engineering, Cairo University. 2005. ISBN: 977-403-011-7

11. Ammaning, Katrin. Analytical Thinking: Why You Need It and How to Get Better. Udemy. 2014. https://blog.udemy.com/analytical-thinking/

12. Systemic Thinking. The Fractal Phenomemon. Systemic Thinking. 2018. http://systemicthinking.com/the-fractal-phenomenon.html

13. Systemic Thinking. How to build a GPS? Systemic Thinking. 2018. http://systemicthinking.com/strategy-fractals/gps/building-a-gps.html

14. Dash, Mike. Critical Thinking. What it is. How it works. Why it matters. Macat.

2012. https://static1.squarespace.com/static/59b7cd299f74569a3aaf0496/t/5afeb67003ce6445740c9563/1526642291284/CT_ebook_Feb_18.pdf

15. Watanabe-Crockett, Lee. Here Are Some Critical Thinking Exercises That Will Blow Your Learners' Minds. Global Digital Citizen. 2017. https://globaldigitalcitizen.org/critical-thinking-exercises-blow-students-minds

16. Nunez, Kirsten. 5 Proven Ways Creativity Is Good for Your Health. Verilymag. 2016. https://verilymag.com/2016/01/mental-emotional-health-creativity-happiness

17. Pennington, Mike. The Top 15 Errors in Reasoning. Pennington Publishing. 2009. http://blog.penningtonpublishing.com/reading/the-top-15-errors-in-reasoning/

18. MDC. Common Fallacies (Or Errors) In Reasoning. MDC. 2018. http://www.mdc.edu/kendall/collegeprep/documents2/COMMON%20FALLACIESrev.pdf

19. *Letter to Henslow*, May 1860 in Darwin 1903. https://www.darwinproject.ac.uk/letter/DCP-LETT-2791.xml

20. Stanford Encyclopedia of Philosophy. Analogy and Analogical Reasoning. Lembeck 1989: 11; Reynolds and Randall 1975: 273. 2013. https://plato.stanford.edu/entries/reasoning-analogy/

21. Stanford Encyclopedia of Philosophy. Analogy and Analogical Reasoning.

Stanford Encyclopedia of Philosophy. 2013.

https://plato.stanford.edu/entries/reasoning-analogy/

22. Balingkang. Analogy and Analogical Reasoning. Scribd. 2018. https://www.scribd.com/document/195269015/Analogy-and-Analogical-Reasoning

23. Learning Express. 501 Word Analogy Questions. Learning Express LLC. 2002. ISBN 1-57685-422-1. https://elearning.shisu.edu.cn/pluginfile.php/36509/mod_resource/content/1/ANALOGIES.pdf

24. English For Everyone. Analogies. English For Everyone. 2018. http://www.englishforeveryone.org/PDFs/Level_12_Analogies_1.pdf

25. Zandbergen, Paul. Statistical Analysis: Using Data to Find Trends and Examine Relationships. Study. 2018. https://study.com/academy/lesson/statistical-analysis-using-data-to-find-trends-and-examine-relationships.html

26. Bakuni, Usha. Types of Statistical Analysis. Study. 2018. https://study.com/academy/lesson/types-of-statistical-analysis.html

27. Research Rundowns. Quantitative Methods, Significance Testing (t-tests). Research Rundowns. 2018. https://researchrundowns.com/quantitative-methods/significance-testing/

28. System Analysis and Design for the Global Enterprise by Lonnie D. Bentley p.160 7th edition

29. Meadows, Donella. Thinking in Systems. Chelsea Green Publishing. 2008.

30. Irny, S.I. and Rose, A.A. (2005) "Designing a Strategic Information Systems Planning Methodology for Malaysian Institutes of Higher Learning (isp- ipta), Issues in Information System, Volume VI, No. 1, 2005.

31. Merriam Webster Dictionary. Methodology. Retrieved: June 2019. https://www.merriam-webster.com/dictionary/methodology

32. Bührs, Ton; Bartlett, Robert V. (1993). Environmental Policy in New Zealand. The Politics of Clean and Green. Oxford University Press. ISBN 0-19-558284-5.

33. Salamon, M.Lester (2002). "The New Governance and the Tools of Public Action: An Introduction", 'The Tools of Government – A guide to the new governance'.

34. Kirst-Ashman, Karen K. (January 1, 2016). Introduction to Social Work & Social Welfare: Critical Thinking Perspectives. Empowerment Series. Cengage Learning. pp. 234–236.

35. Jillian Jimenez; Eileen Mayers Pasztor; Ruth M. Chambers; Cheryl Pearlman Fujii (2014). Social Policy and Social Change: Toward the Creation of Social and Economic Justice. SAGE Publications. pp. 25–28. ISBN 978-1-4833-2415-9.

Printed in Poland
by Amazon Fulfillment
Poland Sp. z o.o., Wrocław

65817745R00113